Edible and Medicinal Plants of the Southern Rockies

Foothills to Alpine in Colorado, Wyoming, Utah and Idaho

By Mary O'Brien and Karen Vail

Copyright 2015 and 2016 by Mary O'Brien and Karen Vail
ISBN 978-1-937862-93-0   **Third printing 2024**
All rights reserved. Printed in China.

United States Library of Congress Number 2015935949

Published and distributed by *Leaning Tree Tales*

All rights reserved. No portion of this book may be reproduced, stored in a retrieval system, or transmitted in any form or by any means – electronic, mechanical, photocopy, recording, scanning, or other – except for brief quotations in critical reviews or articles, without the prior written permission of the author.

Disclaimer

This book is intended as an informational guide, not a how-to reference for using the plants. While many of the plants are edible and/or medicinal, some are poisonous or harmful. This book provides only limited information on their uses. It is not intended to be a substitute for proper training in plant identification. Consultation with a health care professional before using the plants as medicine to treat a health condition is recommended. To the best of our knowledge, the recipes are safe and nutritious. For those people with food or other allergies, or who have health issues, please review the ingredients carefully to determine whether or not they might create a problem for you. All recipes and edible and medicinal uses are at the risk of the consumer.

- Photos by authors Karen Vail and Mary O'Brien
- Book Design by authors and Joan Zinn
- Formatting, layout, editing and some photos by Joan Zinn
  www.medicinehillherbs.com
- Milkweed fruit and seed pod photos by Jim Gibson
- Cover Design by Element Design
- Print-ready production and publishing by BookCrafters, http://bookcrafters.net

Preface to Third Edition

The botanical world classification system has changed significantly since this book was first published. All aspects, from family to genus and species, have changed for many plants. We decided to keep our original classification from reference we used in 2015 noted in our Resources and Webliography. Many web sites have kept up to date on classification and, if interested, please check out these sites for the most recent changes. A noted resource for Colorado classification is now Jennifer Ackerfield's *"Flora of Colorado; Second Edition"* 2023.

Contents

Acknowledgements

Map of the Rockies 1

Pictorial Guide 2

■ Introduction 8

■ Trees & Shrubs 22

■ Herbaceous & Non-Flowering 76

■ Poisonous Plants 212

Medicinal Preparations Appendix 246

Edible Preparations Appendix 250

Illustrated Glossary 254

Glossary of Botanical Terms 256

Glossary of Medical Terms 258

Resources 262

Index 266

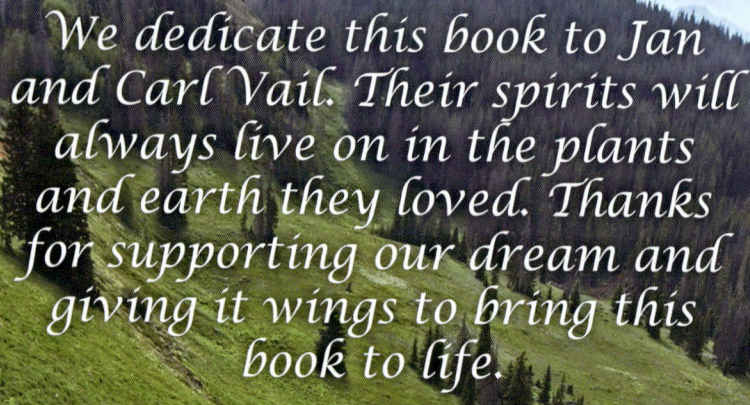

We dedicate this book to Jan and Carl Vail. Their spirits will always live on in the plants and earth they loved. Thanks for supporting our dream and giving it wings to bring this book to life.

Karen and Mary

Thank you Joan Zinn for your creativity, plant knowledge and keen eye for detail!! You were a tireless editor and formatter in spite of our neophyte computer skills.

A special thanks to Mike Kintgen for your encouragement and help in identifying the plants that "stumped" us.

Thanks also to our friends and family and supportive community for your encouragement and PATIENCE in this multi-year process. You all are our rocks!

Rabbit Ears, east of Steamboat Springs, Colorado

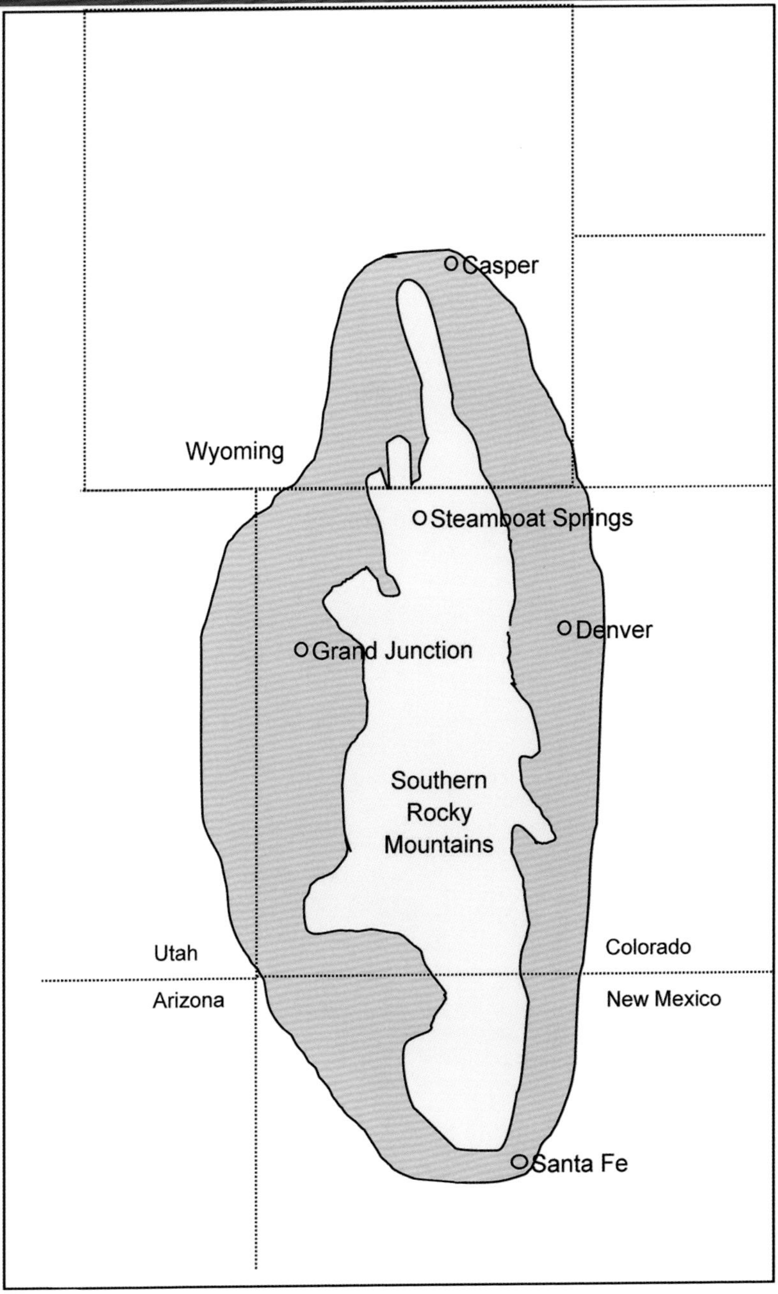

This map is not to scale and placement is approximate.

 Highlighted areas covered in this book

 Southern Rocky Mountains

Pictorial Guide
TREES & SHRUBS

Common Juniper P. 22
Rocky Mountain Juniper P. 22
Subalpine Fir P. 24
Engelmann & Blue Spruces P. 26
Lodgepole Pine P. 28
Rocky Mountain Douglas Fir P. 30 & 32
Rabbitbrush P. 34
Sagebrushes P. 36
Oregon Grape P. 38
Red-Osier Dogwood P. 40
Buffaloberry P. 42
Kinnikinnick P. 44
Huckleberries P. 46
Gambel Oak P. 48 & 50
Golden Currant P. 52
Gooseberry P. 54
Elkbrush P. 56
Serviceberry P. 58
Hawthorn P. 60
Hawthorn P. 60
Chokecherry P. 62
Wild Rose P. 64
Thimbleberry P. 66
Raspberry P. 68
Mountain Ash P. 70
Quaking Aspen P. 72
Willows P. 74

Pictorial Guide
HERBACEOUS

Pictorial Guide

HERBACEOUS

Pictorial Guide

HERBACEOUS

Potentilla P. 188

Strawberry P. 190

Northern Bedstraw P. 192

Figwort P. 194

Mullein P. 196

Broad-Leaved Cattail P. 198 & 200

Stinging Nettle P. 202

Valerians P. 204

Violets P. 206

Horsetails P. 208

Usnea P. 210

POISONOUS

Poison Ivy P. 214

Dogbane P. 214

Poison Hemlock P. 216

Water Hemlock P. 216

Arnicas P. 218

Ragworts P. 218

5

Pictorial Guide
POISONOUS

Red Elderberry P. 220

Mountain Snowberry P. 220

Western Stoneseed P. 222

Mountain Goldenbanner P. 222

American Vetch P. 224

White-Flowered Peavine P. 224

Locoweeds P. 226

Milk Vetches P. 226

Lupines P. 228

Monkshood P. 230

Baneberry P. 230

Columbines P. 232

Larkspurs P. 234

Rocky Mountain Iris P. 236

Death Camases P. 238

False Hellebore P. 238

Paintbrushes P. 240

Buttercups P. 242

Pasqueflower P. 242

Clematis P. 244

Windflower P. 244

Pictorial Guide
KEY TO ICONS

NATIVE NON-NATIVE CAUTION PLANTS ON THE AT RISK LIST PLANTS ON THE TO WATCH LIST

CATKIN FLOWER FORM BERRY FLOWER

"The light died in the low clouds...
Shrouded in silence, the branches wrapped me in their peace.
When the boundaries were erased, once again the wonder: that I exist."

— Dag Hammarskjold

Introduction

THE UTES

Native people have occupied Colorado and Utah for over 8,000 years. The state of Utah's name is derived from "Ute" which means "high land" or "land of the sun." At the time of Euro-American contact, there were 12 different Ute bands inhabiting most of Utah and western Colorado. Though they occupied separate areas, there was frequent intermarriage, trade and regular gatherings of the different bands. They named themselves by the features of the areas in which they traveled and the foods upon which they depended. For hundreds of years the Yamparika band (Yampatika), later known as the White River Utes, visited northern Colorado and southern Wyoming along the Yampa and White rivers for summer hunting and food gathering. They depended on the roots of the yampah plant as a staple food and called themselves the Yamparika which means "carrot or root-eaters." The Uncompahgre Utes (Parianuche or Grand River Utes) lived along the Colorado River (formerly the Grand River). Today these bands along with the Uinta Utes (Utah Utes) are known as the Northern Utes and live on the Uintah and Ouray Reservation located in Fort Duchesne, Utah.

Our focus is on the information available on the Yamparika Utes who were the primary tribe to occupy the area covered in this book. They were primarily hunters and gatherers traveling over their range following the seasons and food supplies. The men hunted deer, pronghorn, buffalo, rabbits, and other small mammals and birds. Women were responsible for the gathering of seeds, grasses, roots, nuts and berries and green plant materials for food, medicine and utilitarian uses. They processed and stored the plant materials and meat for winter use. Chokecherries and other berries were mashed and sun-dried. Sunflower seeds were often ground and cooked and then dried for storage. Other seeds were gathered in finely woven winnowing baskets onto which they would beat the plant to glean the seeds. Roots were dug with pointed sticks three to four feet long. Many of the plants and seeds were dried and placed in baskets that were stored in pits in the ground and covered with earth.

Major plants utilized by the Yamparika, many of which are covered in this book, include:
Food:
    Seeds: from various grasses, sunflowers, amaranth, lamb's quarter.

    Berries: chokecherries, buffaloberries, serviceberries, gooseberries, currants, Oregon grape, huckleberries, strawberries, raspberries, juniper, rose hips and fragrant sumac.

    Roots and shoots: spring beauty, onions, mariposa lily, yellow pond-lily, arrowleaf balsamroot.

    General food plants: cattails, yucca, ponderosa pine bark, arrowleaf balsamroot, dandelion, acorn, pinyon, cow parsnip, biscuit root and numerous greens.

Medicine: Porter's lovage, yarrow, chokecherry, bitterbrush, sagebrush, juniper, cow parsnip, kinnikinnick, Oregon grape, grindelia, and biscuit root.

Introduction

Utilitarian: aspen, juniper, willow, bulrushes, cattails, all evergreen trees, sagebrush, yucca, fragrant sumac, red-osier dogwood, kinnikinnick, chokecherry, serviceberry and other berry woods, and wild rose.

Though little is known about specific Ute management of their most important plants, it is likely their management practices were similar to many western tribes. Yampah (*Perideridia gairdneri*) was highly esteemed by many western tribes and considered a staple food and "were dug by the sackfuls before the blossoms come and kept on hand for winter use." (*Tending the Wild*, p. 296). So common was the use of the digging stick among Great Basin and California tribes that they were called (derogatorily) "diggers" (*Ethnographic Overview*, p. 52). The yampah roots were baked in an earthen oven, then dried, ground on a metate (mealing stone) and stored for winter. Research now has shown us that this plant responds well to digging and has been genetically influenced by human harvesting to produce more tuberous roots than would be produced without harvesting. Yampah was also routinely cultivated by burning patches. Experimental research indicates burning may increase plant density by 10 times or more. Yampah often has branching, spindle-shaped tuberous roots. In digging, these tubers break at the thinnest and weakest point. The remaining tuberous fragments are often composed of both root and stem tissue. By gathering these subterranean parts before flowering and breaking these off to leave pieces behind, humans may have favored those tubers that leave the largest number of fragments (*Tending the Wild*, p. 303). Plant breeder and botanist Luther Burbank reported patches of yampah "growing like grass, so that hardly a shovelful of dirt can be turned over without exposing numerous roots" (*Tending the Wild*, p. 241). Patches of Parsley Family species were also burned at appropriate times to maintain optimal habitat conditions for growth. Chokecherries were also fire pruned to improve fruiting and quality. Dense patches of yampah, glacier lily and Porter's lovage are found in the Rockies and may represent formerly wild-cultivated stands of these species.

Yampah Flower

Willow was a primary utilitarian plant among Western tribes. Shoots were gathered in the fall, winter or early spring. Willow patches were carefully tended by weavers who retuned year after year and cut the plants to the ground after gathering to ensure a harvest of useable shoots the following year. Long, straight shoots, suitable for use in making baskets, proliferated under annual pruning. The willow was used also for sweat lodges, wickiups, dams to catch fish, cradleboards, green dye, sleeping mats, baskets and hats. There is increasing knowledge about the indigenous management of the native plants and ecosystems they depended upon. They developed harvesting techniques to maximize the health and yields of the plants. Many plants were managed through burning, weeding, digging and pruning.

Introduction

How To Use This Book

The plants in this book are separated into 3 sections: trees and shrubs, herbaceous plants (with one non-flowering plant), and toxic plants. The colored banners at the top of each page define each section. Plants are arranged according to their family in each of these sections. Many single species are represented, but if several species have similar uses they have been grouped together. The colored tabs along the page edge indicate flower and/or fruit color or lack thereof.

Plant Names

Common and scientific names follow the taxonomy of the following authorities: Integrated Taxonomic Information System, USDA Plants Database, and Weber and Wittmann's *Colorado Flora: Western Slope,* 4th edition (see **Resources**). There is disagreement in the scientific community regarding taxonomy. In certain cases we have preferred alternative scientific names over currently approved names. Systematics is always changing due to updated DNA evidence and/or re-evaluation of original nomenclature. We have noted alternate scientific names in parenthesis. Other common names are added as well.

Description

A written description, as well as photos, will help you in identification. Remember that this is not intended as a field guide, so if you are ever unsure of a plant, consult other field guides listed in the **Resources** section to make a positive identification. A general description of the plant, including size, flower, leaf, fruit, habitat and flowering times are included, as well as any other appropriate factors aiding identification. Many of the identifying traits of plants vary greatly according to elevation and conditions so our descriptions should be considered as a general guideline.

The best way to become familiar with all stages of a plant is to observe the plant from spring to fall. Observe these changes as the spring shoots emerge, leaves erupt, flowers form, berries ripen, and leaves wither. Because so many plants are used at very specific times of the season, knowing exactly what they look like at each stage by following them through a complete season is invaluable.

One of the first questions we ask when a person asks us to identify a plant is, "Where is it growing?" Habitat is a vital key to plant identification. Most plants, except for the more prolific and adaptable weeds, are pretty picky about where they want to make their home. Once you begin to learn about plants and their habitats, you will anticipate them in these areas. Plants also tell you much about other aspects of the landscape: wet or dry, saline soils, acid soils, rocky soils, cooler or hotter, etc. It is fascinating to be able to read the landscape through the plant life found there.

The **Life Zones** specific to our area characterize a region of more or less uniform climate in which distinctive types of vegetation occur. These zones vary by slope aspect, with north-facing slopes consistently moister and cooler and south-facing slopes being warmer and drier.

Introduction

The **Foothills zone** is characterized by mountain shrublands with a mix of Gambel oak, serviceberry and chokecherry, and higher elevation mountain sagebrush. The elevation ranges from around 6,000 to 8,000 feet. Upper elevations of the Foothills zone and north-facing slopes could see stands of lodgepole pine and Douglas fir mixed with aspen.

The **Montane zone** occurs between 8,000 and 9,500 feet and typically consists of lodgepole pine and Douglas fir mixed with aspen. Engelmann spruce and subalpine fir will be found throughout, especially in the higher elevations and north-facing slopes. Mountain meadows and grasslands occur in moister areas with rich soils. Numerous waterways cut through the landscape creating riparian forests and shrub communities.

The **Subalpine zone** is found between 9,500 and 11,400 feet and is lush with forests of Engelmann spruce and subalpine fir. Limber pine can be found on exposed, rocky slopes. Aspen and lodgepole pine can be found in disturbed areas. Lush meadows, ponds and lakes, and wetlands are prevalent throughout the landscape. The upper reaches merge into the forest-tundra ecotone (where different ecosystems merge), and where the majestic spruce fir forest becomes small and stunted, often forming dense evergreen patches called krummholz forests.

The **Alpine zone** begins at treelimit, around 10,500 to 11,000 feet in our area, and is the treeless expanses of herbaceous plants and dwarf willows adapted to some of the harshest conditions of brutal cold, winds, intense sunlight, rocky soils and short growing seasons.

Introduction

Medicinal Uses

In the plant listings, Mary has written the "Medicinal Uses" section.

Man has been using plants for food and medicine throughout all of history. Through experimentation, animal observation, oral tradition, direct plant communication and ritual, humans have learned which plants and/or plant parts did or did not provide healing powers, or were potentially toxic. I find it interesting that the same plants found growing in different parts of the world often have a history of being used in similar ways.

Plants used for medicine fall into two general categories. The first are "tonic" plants that are closer to food in their actions. They are very nutritional, high in vitamins and minerals. They can be consumed on a regular basis and help to build healthy tissues, bones and natural defenses. It is said we rebuild a new body every 6 months as old cells in each system are replaced by new

Dandelion

ones. These nutritious plants work by replacing weak sick cells with healthy ones — a process that takes time as our bodies constantly regenerate. These are often common weedy plants like stinging nettles, alfalfa, dandelion, red clover and members of the Mustard and Mint Families. These are the safest plants to begin with as they have little or no toxicity.

The second category consists of plants that are more specific in their actions and often affect a specific body organ or system. These are best used only when needed and usually for short term duration, like echinacea used for onset of colds and flu or uva ursi used for kidney infections. The chemical constituents of the plants can tell us the action of the plant: e.g., tannins tighten and tone tissue, bitters stimulate digestion, astringents cause constriction of tissues to stop bleeding, secretions or surface inflammation, and demulcents soothe internal membranes. Why some plants act on specific parts of the body is not well understood by science (it possibly involves constituents that bind to and have affinities for certain cell or hormone receptor sites), but centuries of use has shown us the efficacy of these healing actions. The strongest acting herbs should be used only when needed and ideally under the supervision of an experienced practitioner. Plants run the gamut from gentle and safe to deadly and these actions can also vary among the plant parts. Historical records of plant use do not necessarily imply it was effective; there were probably many failed experiments. This is why personal research, education and proper identification are critical before using any plants.

Arnica

Introduction

I have covered many medicinal plants in this book but I have not used all of them. Personally, I want to learn about all of them on some level and that is a lifetime process. There are many that I gather each year and use regularly and many more that I am excited to learn their names, history and appreciate their gifts, but I am not called to use them. To really "know" a plant —become intimate with it— grow, gather and use it. My favorites and most used are: arnica, willow, aspen, stinging nettle, raspberry, arrowleaf balsamroot, the louseworts, Porter's lovage, wild rose, the wild mints, hawthorn, yarrow, Oregon grape, plantain, dandelion, the horsetails, the violets, usnea, yellow dock, gumweed, valerian and the numerous berries.

Oregon Grape

Much of our current knowledge of plant usage comes from many sources: ancient records (some dating from as early as 5,000 BCE), observation of modern day indigenous peoples, interviews and ethnobotany databases, settler accounts, folklore and folk herbalists' accounts, as well as centuries of empirical data and modern scientific research. Some indigenous communications have not always been considered reliable since it came to us second-hand. The person recording the information may not have been familiar with plants or healing, and strained relations might have caused the informant to alter the information. What many of these sources do not include are specifics of how and when plant remedies were used and their actual effectiveness. In this book, I share information gathered from many of these sources because I am intensely curious and interested in the connection early peoples had with the plants and their intimate relationships with the natural world. This data is supplemented by my own knowledge from personal experience as a practicing herbalist, modern research and current accepted herbal practices.

Efficacy of medicinal plants can be affected by its habitat and age. The active constituents are often concentrated in the young parts of some plants while other plants may be stronger as they mature. Some plants can concentrate stronger constituents in roots or seeds, while the leaves and flowers of the same plant might have a gentler medicine. Many of the old herbals recommend using the strong plant parts, like the roots. These herbals were written when there was no concern about habitat loss or plant extinction and almost every home had a medicinal garden of herbs and spices. Today we cannot be so cavalier. Harvesting the leaves of a plant instead of its roots provides a good medicine without killing the plant. This returns us to the importance of learning about the plants you wish to use, getting to know how they grow, and becoming acutely aware of plants that are being overharvested or threatened in your area. **Gathering Guidelines and Ethics** are found later in the Introduction.

Introduction

Historically the plants were used for medicine by being eaten, prepared into teas (infusions and decoctions), or fermented in vinegars or drinks, like herbal beers or honey-based mead. Plant material was often ground between rocks or with a mortar and pestle. Water, heat and alcohol are universal solvents that make available the plant constituents. Eating and drinking the herb is the most efficient way for the body to absorb the medicine. Teas can be also be used externally in the bath and for compresses and washes for wound and skin care. The plant material itself can be used externally in poultices to treat injuries.

Teas, tinctures, capsules, salves and creams are easy to make in the home kitchen. Making your own remedies can increase the energetic healing power in the remedy and expand your understanding and connection to the plant itself. Today, it is common for herbal remedies to be found in commerce as capsules and alcohol extracts (tinctures) which offer convenience and longer shelf life. Tinctures are readily absorbed by the body. They are usually taken diluted in a little water or juice. Capsules must be broken down in the digestive system in order to be absorbed and this sometimes interferes with their efficacy. Also, to be encapsulated the plant material must be reduced to very small particles which can shorten the shelf life. Be sure to purchase the freshest capsules available (check expiration dates).

The **Herbal Preparations Appendix** offers brief descriptions of various preparations. Specific preparations are not covered in detail in this book; see the **Resources** section for a complete list of other resources.

Yarrow

Growing your own medicinal plants is a powerful way to create connection to their healing powers. By growing native and native-adapted medicinals you are helping to maintain their ecological viability and reduce over-harvesting from the wild. Many traditional, well-known and cultivated garden herbs used for cooking have also been used medicinally throughout the ages. Consider adding perennials like thyme, sage, oregano, hyssop, lavender, lemon balm, garden mints and also members of the Parsley Family (*Apiaceae*) to your medicinal garden plantings.

The information in this book is not intended to become a basis for a home pharmacy, but instead to introduce the reader to the plants' historical uses and potential as therapeutic allies. No one book can cover all you need to know to put wild plants to medicinal use. I recommend you thoroughly research any plant you consider using through all mediums available and preferably get hands-on training with a qualified professional. Please seek the care and advice of a qualified medical practitioner for all medical problems. If you are pregnant, have a chronic condition or take medications, it is best to consult a health-care professional before using home cures.

Introduction

Edible Uses

In the plant listings, Karen has written the "Edible Uses" section.

There are an estimated 31,000 plus kinds of plants in North America (which in this case includes North America north of Mexico plus Hawaii and Greenland; strange collection, but this is how the scientists break it out) according to Daniel E. Moerman's *Native American Food Plants: An Ethnobotanical Dictionary*. Mr. Moerman states that the Native Americans used more than 1,500 plants for food and this number is probably very conservative. If I were to count the numbers of plants offered in our local supermarket, I know the numbers would pale in comparison to the rich diversity available to our earlier ancestors. I would love to look past just the numbers to a lifestyle that was interwoven with every aspect of the natural world. These people, who were totally dependent on the land for all their needs, recognized every nuance of

Thimbleberry

the seasons, the plants and animals they harvested, and the landscapes they lived in and moved through. The knowledge that has been lost as we have moved into a very different relationship with nature is disheartening, to say the least. Our connection to the Earth can be strengthened by our understanding of plants in such an intimate way as we harvest them respectfully for our use, similar to the Native American reverence for life.

There are many ways to eat plants. My favorite ways of preparing plants are found in the **Edible Preparations Appendix**. I have tried to keep the uses simple. In most cases I have personally tried the plant in the way I have written it up. If I did not have personal experience with the plant, I have double checked the uses with several references to be sure of its accuracy. It is amazing as I was going through these references how often they disagreed on the part used or when to gather, etc., so please take that into consideration. That said, it is imperative that, as with any new food you try, you do so in stages and make sure you are listening to your body's reaction. Introduce the new food into your diet slowly and in small portions. I will never forget getting a call from a lady who had come on one of my walks and was very taken with the glacier lily pods that taste like green beans. I had added the caveat that they should only be consumed in moderation as they are emetic. Hence the call after she added a huge handful to their salad that evening. Also, as Mary and I were writing this together, we sometimes had very different experiences with the plants. Some that I thought were quite tasty were only marginal for her. Neither of us is "right" or "wrong" in how to enjoy and use the plant; we all have different tastes and preferred ways of preparing plants, as did our earlier ancestors. Find your favorites. Our present day diets of high sugars and salts may taint the palatability of some plants. There are a few recipes throughout the book for you to try. Check the **Resources** section for many more recipe books.

If I were to pick 10 of my favorite edibles I would choose yampah (young leaves, flowers and seeds — I harvest the root from my garden when I really need a root fix), serviceberry (in

Introduction

good years I fill my freezer and canning shelf with tasty treats to use all winter), cattail (I have been using the pollen more and more and find it a wonderful addition to flours), dandelions (I love the spring leaves in salads, the flower buds in stir fries, and the sunny, open flowers as fritters), sweet anise (because I have a sweet tooth, this naturally sweet plant is a must have for me and grows in my garden to have at hand and grate the root on ice cream, and add seeds, flowers and leaves to salads), onions (our abundance of sweet and spicy onion flowers are a treat), wild roses (the petals are my favorite trail munchie as well as the rind of the ripened hips), the fruits of the Rose Family (raspberries, strawberries, thimbleberries, chokecherries… yummy!), stinging nettle (definitely an acquired taste, but the leaves are so good in soups and stews) and our field mint (a refreshing treat raw or added to other dishes). I am sure you will soon have your own list of favorites from the abundant offerings of our area.

Serviceberry

What should you bring to collect wild edibles? Well, the most important is your brain! Make an absolutely positive ID before even tasting a plant. Take a field guide or knowledgeable friend. Use your senses. I am always smelling plants, even if I know what they are, because that is so ingrained in my ID process. I rarely collect roots any more, unless it is from a plant that I can take a piece from and replant, like sweet anise. But if you are digging, a narrow shovel like an irrigation spade will create the least upheaval in the ground around where you are digging. A trowel can be used for smaller roots. Take scissors or pruning shears to cut leaves, stems and flowers. Trying to break a stem can often result in pulling the whole plant out of the ground. If you are collecting several different items, bring a bag (I love canvas bags) or container for each so they are not mixed. If you are collecting greens, it is best to place them in a waterproof bag so they do not wilt. For berries I love the Gardener's Leg, a nifty container that has a waist belt and a pouch hanging down by your side.

— Go forth and eat! —

Notes

Take yourself back in time when there were no stores to purchase food, medicines, household goods, clothing, leisure time materials, etc. Everything you needed came from the natural world: fibers, basketry, building, tools, fuel, dyes, poisons for hunting and fishing, and toys and recreation. It was imperative that you learned your lessons well on what and how to collect and use items, or you would not survive. The Notes section highlights these other plant uses. Of course, we have modern day uses in there as well.

Wild Rose Hips (fruit)

Introduction

Garden Notes

Garden notes throughout the book provide tips on medicinal and edible plants to include in your own garden. We both believe that native and native-adapted plants should be an integral part of our gardens. They offer more than just the uses described here, they are valuable pollinator plants for local pollinators, are disease resistant, well-adapted to the local climate and tend to be tough as nails in the right situation. By growing native and native-adapted plants, you are helping to maintain their ecological viability and reduce overharvesting from the wild. Growing your own medicinal and edible plants, many of which are found in this book, is a powerful way to create connection to the gifts of the plant world. See the **Resources** section for books on gardening, or visit your local County Extension Agency for a plethora of information. Many local nurseries are offering more and more native and native-adapted plants. Good for them! By supporting their efforts we can encourage even more native and native-adapted plants to be introduced into the nursery trade.

Caution

A plant's goal in life is to regenerate itself through setting seed or by a form of vegetative reproduction. To get to that goal, plants need to avoid being eaten. That, of course, is the goal of many animals to survive — eating plants. Survival strategies for plants range from physical barriers of thorns, hairs, waxes, etc., to chemicals that taste bad or poison the eater, or are irritants. These defenses are not specific to wild animals; they can affect us as well.

Please read and heed notes in the yellow caution boxes about plants, trees and shrubs before using them as a food or medicine. Awareness of any potential dangers, toxicity, or allergic reactions arms you with important knowledge that serves you well in the wild. Proper and certain plant identification is critically important when you are learning to use wild plants as food and medicine without a learned practitioner at your side, guiding your every step of the way. Do not be afraid – be aware, be informed, and proceed with caution.

Introduction

Gathering Guidelines and Ethics

1. The most important item to take when gathering wild edibles and medicinals is your brain! Make an absolutely positive ID before even tasting or using a plant. Take a field guide or knowledgeable friend. Start with familiar plants like common weeds. Study the poisonous plants that grow in your area, and always know whether the plant you're harvesting has any poisonous look-alikes. Some plants have parts that are both edible and toxic, and some plants have parts that are edible at certain times of the year but toxic at others. Make sure you know which part to use and how and when to harvest it.

2. Practice stewardship with respect for nature. Talk to the plants, ask permission, give thanks, make an offering, express gratitude, acknowledge and become part of the connection with all life. Stay present throughout the process and the quality of your food and medicine will reflect your efforts.

3. Ask permission before harvesting on private property. Check the regulations for the public lands you might visit.

4. Observe if the stand can support harvesting. Is it in a healthy ecosystem? Is the plant endangered? Check the United Plant Savers or Colorado National Heritage Program for lists of endangered plants (see **Resources**). How many plants can the plant community survive without, not how many plants you need in order to make products or profit. Harvest conservatively and only what you will truly process and use. Preferably, pick from areas about to be developed or cultivated. Return often to your harvesting sites to get to know the plants at different phases of their lifecycle and watch for impacts of harvesting. Gather from areas you know and love and you will instinctively protect them. If gathering for food, taste the plant first. You might not like it!

5. Are there at least 10 plants? General guidelines say take one plant for every ten or 10% of plant material (leaves, flowers, etc.). Weeds grow fast and abundantly and can be harvested in larger numbers while other plants have been so over-harvested that even if you find 10 growing in an area, harvesting should be discouraged. Native plants are an intricate part of their ecosystem and can grow slowly and should get special consideration when harvesting — some guidelines say no more than 5%. All this emphasizes that it is important to get to know the individual plant in order to harvest it sustainably.

6. Leave the grandparent plants (these are the mature and seed producing plants within the stand) and plants at the top of a hill to seed downslope. Pick healthy, unstressed plants, no yellow leaves, bug holes, etc. Do not girdle trees when taking bark — it will kill the tree. When gathering medicinal plants, harvest in the early morning after the dew is off leaves and flowers: 8-10 a.m. is the ideal. Avoid picking on rainy days as there is too much moisture in the plant material.

7. Are there possible contaminants in the soil? Avoid areas that are likely to be sprayed, for example around power lines, train tracks, golf courses, and weedless yards. Don't gather

Introduction

within 50 feet from a busy road, especially downhill from one. Don't gather on hiking trails or near popular attractions and other high-use areas. These areas have higher contamination potential (think dogs) and are also valued for what they offer in terms of beauty and uniqueness. Do not trample areas outside of designated trails, or disturb the environment by over-collecting from a single species. Look for signs of other impacts.

8. The time to harvest depends on what plant part you are using and where the plant's energy is concentrated.

    - Early spring/summer: young leaves, catkins, bark. Some leaves can be harvested throughout the season from plants that continue to produce new leaves: e.g., dandelion and raspberry.

    - Summer: flowers, when beginning to bloom.

    - Late summer/fall: bark, seeds, fruit. Replant some seeds.

    - Fall: roots after plant growth cycle is complete and energy has moved back down into roots. Replant root crown if possible to propagate new plants and sprinkle seeds. But first, ask yourself whether you really need the roots. Many plants have the same properties in their aerial portions. For other plants, aerial portions can be combined with roots to make whole plant medicine, using less root material.

   Moon phase affects plant energy and tincturing: new moon to full moon is the traditional phase for harvesting and tincturing.

9. Start a replanting project in your area to help reestablish endangered and threatened species. Many native and native-adapted plants do well with cultivation. Use your garden as a source of edibles and medicinals. This lessens our impact and helps preserve the species.

10. Respect the wildlife of the area. Their survival depends on the roots, shoots and seeds you may be harvesting. If collecting from shrubs and trees, the Utes would collect at chest height to leave the upper reaches for birds and the lower areas for the little critters.

Glacier Lily

Introduction

Our Natural World Is Changing

A report from the Union of Concerned Scientists and the Rocky Mountain Climate Organization on September 10, 2014 (*Rocky Mountain Forests at Risk: Confronting Climate-driven Impacts from Insects, Wildfires, Heat, and Drought*) provides undisputed evidence that man-made climate change is altering our forest, weather, waterways and phenology (the timing of a periodic biological phenomenon in relation to climatic conditions). According to this report, if current climate trends continue, we could have a variety of impacts: bark beetle infestations will spread, large, intense and more frequent fires will destroy more acres (even if temperatures rise only a little), less snow will cover the ground in the spring, the snow will melt sooner, running down streams before trees can suck it up, and several tree species such as lodgepole pine, Engelmann spruce, ponderosa pine, Douglas fir, whitebark pine, aspen and pinyon pine could likely go extinct. There are many more excellent studies out there looking at a host of climate change driven changes: earlier flowering times that don't mesh with the normal pollinators, droughts that are decreasing disease resistance in aspen and many conifers, frost damaged buds of subalpine species leading to fewer flowers and subsequent seed production, deposition of dark dust (blown in from drought ridden deserts) on winter snowpacks hastening melting and increases in weedy species in native habitats. That is a long list of rather frightening scenarios, many of them already occurring in our world.

By just perusing this book, this shows that you have an interest in our natural world, and, as a forager, your connection runs deep. There is a lot of science behind climate change, and that science provides a sound basis for action. But the real call to action comes from deep within our connection to the earth. Our actions, whether they be personal decisions, educating the public, political or otherwise, are what will mark our future path, and we are in dire need of some grounded thinkers and doers to walk this path. Plants are the basis for all life on earth. Your relationship to them provides a strong connection and base with which to call upon. We cannot sit back and wait for the earth to heal this time around. We have pushed too far. Now we need to all work diligently to negate the doomsday scenarios stated earlier and make a conscious effort to walk a path of healing.

> "The earth may be borrowed but not bought. It may be used but not owned. We are the tenants and not possessors, lovers and not masters. The Earth belongs to the wind and the rain, to the sun and all seasons, to the cosmic secrecy of seed, and beyond all, to time."
>
> — Marjorie Kinnan Rawling

Introduction

Dad and brother Vail fishing at Hahns Peak Lake — catching the BIG ones!

Demystifying Conifer ID

Flat, friendly, phallic fir
Sharp, square, sagging spruce
Pine are in packets

This is the special code for general conifer identification. There are three main types of conifers: spruce, fir and pine. If you come to a conifer and pull a needle off and they are in packets of 2, 3, 5 or 7, you have a type of pine because "pine are in packets." The needles are connected at the base creating a "packet."

Now that you have ruled out a pine in your conifer ID, your two next choices are fir or spruce. Now you need to introduce yourself to the tree, so give the branch a good handshake. If it is soft and friendly, you should take a needle and try and roll it to see if the needle is flat. If you have a friendly, flat needled tree, you have a fir!

Now shake that branch again. If it is sharp (ouch!) and the needle rolls easily between your fingers because it is square, you have a sharp, square spruce!

Now look up at the cones of the spruce or fir and if the cones are sagging (hanging down), you most likely have a spruce. If the cones are phallic (sticking up from the tips of the branches), you will have a fir.

There is an exception for the Douglas fir that has cones that sag, but remember, the Douglas fir is not a true fir but a "false" fir...poor thing!

Cupressaceae Family — Junipers
Juniperus communis var. *montana* & *Juniperus scopulorum*

Rocky Mountain juniper habit and fruit (inset left)

Common juniper habit and fruit (inset)

Cypress Family

Common Juniper

Juniperus communis var. *montana* (*Juniperus communis* ssp. *alpina*)

Rocky Mountain Juniper

Juniperus scopulorum (*Sabina scopulorum*)

Other names: (for Rocky Mountain juniper) red cedar, savin

Description: **Common juniper** is a low growing (up to 4 feet), spreading evergreen shrub with sharp needle-like leaves and small bluish "fruits" (actually fleshy cones). Most common as understory of coniferous forests. **Rocky Mountain juniper** is an upright evergreen shrub to 15 feet with scale-like leaves and small bluish "fruits" (also fleshy cones). Dry, rocky open sites into montane.

Medicinal uses: Juniper is tonic, diaphoretic, carminative, diuretic and strongly antiseptic, especially within the urinary tract. It is used for cystitis and fluid retention. It is settling and warming to the digestive system. It stimulates menstruation and increases menstrual flow. Collect the berries when blue to purple. The berries infused in oil are used in salves for wounds, chronic arthritis and rheumatic pains. A handful of leaves or berries added to warm bathwater can soothe aching muscles. One or two ripe berries chewed daily help prevent infection when exposed to contagious diseases. Chewing 1 or 2 berries before meals can stimulate hydrochloric acid and pepsin production. An infusion of berries and/or leaves as a steam inhalant is helpful for nose, throat and lung congestion.

Edible uses: Collect the blue to purple berries — they will be sweeter; although I love the punch of the green berries, which will stay with you the rest of the day. Eat them raw, boil, roast, or add to wild game dishes. The dried berries can be ground into a meal to make mush or cakes. The berries are distilled to flavor gin. Burn the green leaves, put boiling water on the ashes and use the strained liquid as a flavoring in other foods. The inner bark was considered starvation food by Native peoples.

Notes: Boil the berries, then skim off the top or allow the water to evaporate and collect the fragrant wax. The bark, berries and needles are used for a brown dye. As a mordant, burn green juniper needles, mix ashes with the dye. Beads were made from the berries by placing ripe berries on ant piles. The ants would industriously eat away the flesh, leaving the hardened seed with a hole in it. The berries germinate only after the waxy coating has been removed by passing through the digestive tract of an animal. The wood was used to make bows. Leaves and berries are used in essential oils, smudge, and potpourri. The plant protects from evil spirits and disease.

Garden notes: Both common juniper and Rocky Mountain juniper are now available in the landscape industry. They will do best with very dry conditions in poor soils. Don't expect fast results! They are very slow growing.

Caution: Large doses and long term internal use are not recommended. Pregnant women should avoid this plant internally. Diabetics should avoid long term use due to possible affects on blood glucose levels. Oil of juniper should not be used internally.

Subalpine fir

Pinaceae Family — *Abies lasiocarpa var. lasiocarpa*

Subalpine fir habit

Subalpine fir bark

Subalpine fir needles

Subalpine fir cones

Pine Family

SUBALPINE FIR

Abies lasiocarpa var. *lasiocarpa*
(*Abies bifolia*)

Other names: Rocky Mountain fir, corkbark fir

Description: A spire-like evergreen tree up to 120 feet tall with silver-gray, smooth bark with bulging resin blisters. Subalpine fir is often seen as a multiple-topped tree because its soft wood is easily broken. Single needles usually less than one inch long, blunt tipped, lacking a "foot" at the base, flattened, bluish-green with rows of tiny white dots (stomata) on both sides. The needles tend to turn up, not rotating around the stem. Male and female are cones on the same tree. Seed (female) cones are purple when young in June and July, maturing to a lighter purple cylindrical cone sitting upright on the branch. The scales are shed with the seeds at season's end leaving the slender central core. Found often in mixed forests with Engelmann spruce on mountain slopes into the alpine.

Medicinal uses: I have no experience using our subalpine fir, although other species have been used medicinally in the past that are largely ignored in the herbal medicine world today. The bark decoction was used as a tonic and to treat colds and flu. The gum was softened in water and applied to wounds. The gum tea was used as a cleansing emetic. It was also chewed to treat bad breath. The leaf tea was used as a laxative and to treat coughing up of blood which can indicate TB. A poultice of the leaves was used to treat chest colds and fevers. It is likely that our species can be used similarly.

Edible uses: The young needles are used as a tea substitute. The cones can be ground into a fine powder, then mixed with fat and used as a confection. It is said to be a delicacy and an aid to digestion. The resin from the trunk is used as a chewing gum and was also made into tea. The inner bark is often dried, ground into a powder and then used with cereal flours when making bread. The seeds are very small and fiddly to use. Seeds of this genus are generally oily with a resinous flavor and can be eaten raw or cooked.

Notes: Subalpine fir species probably have similar historical uses. The fragrant young leaves and twigs were used to repel moths or burnt as incense. They were also ground into a powder and used to make a baby powder and perfumes. A gum obtained from the bark, which is antiseptic, was chewed to clean the teeth. It was also used to plug holes in canoes. An infusion of the leaves was used as a hair tonic. The leaves were placed in the shoes as a foot deodorant. The wood was used for making chairs and insect-proof storage boxes. The wood is light, soft and not very strong and is little used except as a long burning fuel and pulp.

Garden notes: Because subalpine firs are such a "soft" tree, they are easily broken by heavy winds and snow loads. Their narrower conical shape allows them to fit into smaller spaces better than the broader spruce trees.

Caution: The wood, sawdust and resins from various species of pine can cause dermatitis in sensitive people. Frequent use or high concentrations of bark or needles internally can irritate the kidneys. Always use evergreen teas in moderation.

Pinaceae Family — Spruces — *Picea spp.*

Engelmann spruce cones

Engelmann spruce bark

Colorado blue spruce needles

Engelmann spruce habit

26

Pine Family

Engelmann Spruce
Picea engelmannii

Colorado Blue Spruce
Picea pungens

Description: **Engelmann spruce** is a stately conical evergreen up to 120 feet tall with thin light gray to brown scaly bark. Single needles are blue-green, 4-sided (roll it between your fingers), sharp-tipped, somewhat curling up at the tips and with a stout "foot" at the base. Male and female are cones on the same tree. Pollen (male) cones are small dark purple to yellow; seed (female) cones are reddish purple when young in June and July, maturing to pendulous papery cones with jagged scale ends which fall intact in winter. Found in cooler canyons and slopes from montane to subalpine. At timberline they are contorted into krummholz (crooked wood) forests. **Colorado blue spruce** has darker, thicker, rigid bark, stiffer, sharper needles, often with a bluish cast, and larger cones with rounded scales. They are found only near water ways.

Medicinal uses: I can find no history of use of the Colorado blue spruce as medicine but the Engelmann spruce has various medicinal uses. The bark tea was used to treat respiratory problems and TB. The gum mixed with fat was used for boils, abscesses, eczema, cuts, scrapes and other skin problems. A decoction of the leaves and gum has been used in the treatment of cancer. It was said that if this treatment did not work then nothing would work. This decoction was also used to treat coughs. Ashes of the twigs, mixed with fat or oil, were used as an ointment on damaged skin.

Edible uses: Spruce tip beer, using the young growing tips, makes a tasty beverage. The young shoots can be used in teas and, because of their high Vitamin C content, were used to help prevent scurvy. The inner bark can be used as an emergency food, and Native Americans would dry and pound the inner bark into a nutritious flour in times of starvation.

Notes: Native peoples used **Engelmann spruce** limbs and roots shredded and pounded to make cord and rope. The boughs were used to line the floors of sweat lodges and camping beds. The bark was used to make baskets, toys and various utensils and is a source of tannin. The lumber is close grained, soft and not strong, used for lumber, fuel, charcoal and to make paper. **Colorado blue spruce** is the state tree of Colorado. The wood is light, weak and full of knots and has little commercial value, but is used in construction and in the pulp industry to make paper.

Garden notes: **Engelmann spruce** are cool, moisture loving trees and enjoy these conditions for optimal growth. **Colorado blue spruce** is used extensively in the landscape trade because of the beautiful silver-blue colors. I have seen far too many wonderful spruces planted then left with no care, where they perished within a couple of years. A slow agonizing death for a special tree. Adequate watering in the first couple of years to help the tree establish a vigorous root system will help secure a long life.

Caution: The wood, sawdust and resins from various species of pine can cause dermatitis in sensitive people. Frequent use or high concentrations of bark or needles internally can irritate the kidneys. Always use evergreen teas in moderation.

Lodgepole pine

Pinaceae Family — *Pinus contorta var. latifolia*

Lodgepole habit

Lodgepole cones

Lodgepole male cone

Caution: The wood, sawdust and resins from various species of pine can cause dermatitis in sensitive people. Frequent use or high concentrations of bark or needles internally can irritate the kidneys. Always use evergreen teas in moderation.

Lodgepole Pine

Pinus contorta var. *latifolia*

Other names: Rocky Mountain lodgepole pine

Description: A tall slender evergreen up to 120 feet tall, most commonly in dense monoculture, even-age stands with little understory plants. Needles are in packets of 2, less than 3 inches long. Cones are woody and bristle-tipped, somewhat asymmetrical and persisting, sometimes for the life of the tree, on the branch. Some cones open when mature, other cones require heat to open and remain as tightly closed cones. Found in montane, typically drier sites. They form dense stands after fires.

Medicinal uses: All pines have similar properties and uses. The antiseptic sap or pitch was used by Native people externally, alone or mixed with other herbs and fat, in a poultice or warmed into a paste for splinters, aching muscles, arthritis, sores, cuts, burns, lung congestion, coughs, colds and sore throats and used in preparations taken internally for many of these same conditions. It has been shown to have antifungal effects in laboratory experiments. Just chewing a small piece of the pitch can create softening and expectoration of bronchial mucus. Needle and bud tea is high in vitamins A and C and used for stomach pain, general debility, coughs, colds and flu. The inner bark was used as a dressing for scalds, burns and skin infections and either eaten or taken in a tea as a cleansing laxative, diuretic and blood purifier as well as for coughs, colds, flu, TB, gonorrhea and stomach aches. Turpentine obtained from the resin of all pine trees is antiseptic, diuretic, rubefacient, vermifuge and vulnerary. Though not used by herbalists much today, it was a valuable remedy when taken internally in the treatment of kidney, bladder and respiratory problems, mucous membrane diseases and externally as a liniment and poultice for various skin problems, wounds and rheumatic complaints.

Edible uses: Pines were, and still are, considered to be food in emergencies. They can be very bitter and sometimes tough to digest. The soft inner bark was used historically by mashing the fiber and making cakes. The cakes were placed between false hellebore (*Veratrum tenuipetalum*) leaves, a fire of moist materials was placed on top and they were baked for an hour or more. They were then smoked, pressed firmly, then stored until needed for longer trips. Needles from any pine make an "acceptable" tea.

Notes: The wood was used for travois, tepee poles and lodges and today is used in the building industry for home framing, furniture and for fuel. The pitch was used as waterproofing and glue. Tan or green dye was obtained from the needles. The roots were braided for rope. Lodgepole pines are fire adapted trees whose cones need exposure to high heat in order to open and release their seeds. They are one of the most widely distributed New World pines and the only conifer native in both Alaska and Mexico. It is a larval host and/or nectar source for the Western Pine Elfin butterfly.

Garden notes: The lodgepole pine, next to the Colorado blue spruce, is probably one of the most widely planted evergreen trees in Northwest Colorado communities. They are tough, drought tolerant trees, with hardly anything that can kill them except a little tiny beetle the size of a grain of rice. The pine bark beetles have decimated the lodgepole pine populations in western Colorado after several years of drought and left the trees weak, making it easy for the beetle to burrow under the bark. A huge "blowdown" north of Steamboat Springs in 1997 left thousands of trees for the beetles to dine on.

Pinaceae Family — Douglas-fir — *Pseudotsuga menziesii*

Douglas fir bark

Douglas fir habit

Douglas fir needle

Pine Family

ROCKY MOUNTAIN DOUGLAS-FIR

Pseudotsuga menziesii var. *glauca*

Description: A beautiful conical evergreen up to 180 feet tall with smooth silvery bark on younger parts of the tree, aging to corky ridges with fissures. Single needles are very soft and thin, flat, about 1 inch long, bright green and fragrant. Pendulous cones are soft with exserted bracts looking like a mouse tail and two legs sticking out. Cones mature in one season. Locally abundant on north facing slopes and canyons and scattered up to the alpine.

Medicinal uses: Various native tribes used this tree to treat a variety of complaints but it is rarely used by herbalists today. The antiseptic resin as a poultice was used to treat cuts, burns, wounds and other skin problems and to treat injured and dislocated bones. Chewed, it was a treatment for sore throats. The inner bark tea was used to treat excessive menstruation, colds, coughs, sore throats, gonorrhea, stomach problems, intestinal pain and constipation. Young needle and bud tea is high in vitamin C, mildly diuretic and expectorant and was used as a general tonic, and for colds, coughs, sore throats, urinary tract problems, venereal disease, scurvy, rheumatism and as an antiseptic wash. The leaf tea was used as a wash for rheumatic and paralyzed joints. The young sprouts or twigs as a tea were used for colds, kidney and bladder problems. The young shoots were placed in the tips of shoes to prevent perspiring and athlete's foot. A mouthwash was made by soaking the shoots in cold water.

Edible uses: Fresh needles are steeped in hot water for a tea high in Vitamin C. The small winged seeds were eaten. The inner bark was dried and pounded into a meal and mixed with cereals for making bread as a famine food.

Notes: Not a true fir but one of the most significant timber species on the west coast of the United States where it can grow several hundred feet high; Intermountain Douglas-fir is smaller, grows more slowly, and has bluish foliage. The bark is a source of tannin, a light brown dye, fertilizer and can be used as a cork substitute. The bark is valued as a fuel because it contains pitch that burns with a lot of heat but very little smoke. The small roots were used to make baskets. The boughs, twigs and shoots were used in sweat lodges and other ceremonies. The resin can be used to make glues, candles, fixative and caulking material on boats. The wood is heavy, strong, fine-grained, and durable, dries quickly and does not warp and is used for heavy construction, telephone poles, furniture and is considered a good quality fuel.

Garden notes: This stunning, sturdy tree has been underutilized in the landscape. I have had a difficult time getting them established, but once they have rooted in well, after about two years with plenty of additional water, they will tolerate basic neglect, little water and high winds. Given a little bit of TLC, they will grow into the majestic trees we enjoy in nature.

Caution: The wood, sawdust and resins from various species of pine can cause dermatitis in sensitive people. Frequent use or high concentrations of bark or needles internally can irritate the kidneys. Always use evergreen teas in moderation.

Pinaceae Family — Douglas-fir — *Pseudotsuga menziesii*

Douglas fir habit

Douglas fir cone

The Legend of the Douglas Fir Cone

Many moons ago, before there were ever human people in the world, there were animal people. They lived in a cold, dark world with very little food. But they did not complain because that is all they had ever known. One day as they were all sitting eating their boiled root mash one asked, "If we could change one thing in our world, what would we change?" They all had a spirited conversation for quite some time. Many wanted more heat so they did not shiver all the time. Many wanted more light so they could see what they were gathering and enjoy the rest of the animal people. Many wanted more food. Eventually everyone agreed that more food would be the wisest. So they called down the Great Spirit and in a trembling voice asked this great provider if he could give them more food. He thought for a while and said, "I can give you more food, but it will not be easy to find. You must search well to discover this new food source."

All the animal people were so excited they could not sleep that night, and in the meager light they called morning they peaked from their wickiups to see if the Great Spirit had left them their new food. "Aah, the Great Spirit said this will not be easy," said Bear. So the animal people began their search. They went to the vast meadows to look, and saw nothing new. They went to the rolling hills to look, but, again, found nothing new. They searched by the small streams and the great river, in the rock piles and in the vast forest. But they found nothing new—no new food that the Great Spirit had promised them. Many of the animal people thought the Great Spirit had forgotten them so they went to get their digging sticks to dig up the bland roots that were their main food. But Squirrel could not be discouraged so easily! She ran through the forest poking under this log and that, then ran up to the top of a dark, tall tree and spotted a brown thing she had never seen before. Excitedly she took the brown thing and pried it open to find small little things inside. She delicately bit down. "This is it!" she cried as she leapt down tree and ran toward the camp of the animal people. "The new food is in the tops of the tall, dark trees!"

Of course the greediest of all the animal people, Mouse, was the first to the trees. By the time the other animal people had arrived Mouse had devoured many of the new brown things. The animal people were furious and called down the Great Spirit. "Have you enjoyed the new food I gave you?" asked the Great Spirit as he appeared. The animal people bowed their heads and pointed to the top of the tall, dark tree where Mouse was still devouring the brown things. "Mouse has eaten almost everything you have provided." declared Moose. The Great Spirit looked into the tree, frowned, and then suddenly clapped his hands with an ear-shattering bang. At that instant the brown things curled up around Mouse, holding his body inside but leaving his tail and two hind feet hanging out. To this day Mouse is still visible hanging from the Douglas fir cones to remind us all to not be greedy.

- Author unknown

Asteraceae Family | Rabbitbrushes | *Chrysothamnus & Ericameria* spp.

Rubber rabbitbrush habit and rounded flowerhead clusters (above)

Yellow rabbitbrush flowerheads (left oval) and habit (above)

Caution: This plant has been reported to be toxic to livestock; however, the plant is so unpalatable that quantities sufficient to cause toxicity are not likely to be ingested. Some people are allergic to the sweet-peppery scent of the pollen of yellow and rubber rabbitbrushes.

Habit and elongated flowerhead of Parry's rabbitbrush

Sunflower Family

RABBITBRUSHES
Chrysothamnus and Ericameria spp.

Description: The two more common higher elevation rabbitbrushes are yellow rabbitbrush and Parry's rabbitbrush with rubber rabbitbrush found in lower elevations of the foothills. They all have similar habitats of very dry, often disturbed sites. All bloom very late in the summer, August to September, heralding summer out in a glory of golden yellow. All our rabbitbrushes have small narrow leaves, small yellow tubular flowers lacking rays arranged in terminal and showy clusters.

Yellow rabbitbrush (Dwarf rabbitbrush, green rabbitbrush, sticky rabbitbrush) (*C. viscidiflorus* ssp. *viscidiflorus*) is a lower (up to 3 feet), rounded shrub with glabrous linear leaves, typically twisted. The yellow flowers are sticky (hence the Latin *"viscidiflorus"* meaning "sticky flowers").

Parry's rabbitbrush (Mountain rabbitbrush) (*E. parryi* var. *parryi*) is a bright green ragged looking shrub up to 2 feet tall with long, narrow leaves sometimes spiraling and larger flowers than other rabbitbrushes. The phyllaries are light green, narrow and keeled, i.e., humped or raised like a ship's keel.

Rubber rabbitbrush (Common rabbitbrush, chamisa, chamiso blanco, gray rabbitbrush, goldenbush, false goldenrod) *(E. nauseosa* var. *nauseosa)* can grow to 5 feet into dense shrubs with silvery-gray leaves and numerous small flowers in broad clusters. It can reseed heavily.

Medicinal uses: Native and early settlers used the leaf in decoctions to treat toothaches, fevers, coughs, TB, chest pains, venereal diseases, stomach cramps and constipation. A strong tea was added to bath water to reduce fevers and the swelling and pain of arthritis and used as a wash for sores and skin eruptions like chicken pox, measles and small pox. The root decoction was used to treat coughs, colds, fevers and menstrual cramps. The mashed leaves were packed into decayed teeth to relieve toothaches. Research shows some antibacterial activity. I know of herbalists today that use rabbitbrushes in ointments and salves.

Edible uses: The milky sap from the bark and root was used as a chewing gum. Lumps and knots on the limbs were chewed for gum.

Notes: The milky sap is a small commercial source of latex, and was studied extensively during World War II as a substitute for commercial rubber. A green dye is obtained from twigs and bark and a yellow dye from the flowers. The branches were slowly burned to tan hides. The leafy boughs were used to cover sweathouses and carpet the floor and for fiber to make mats, building materials and for weaving. The plant infusion or smoke from burning the plant was used to relieve fright and keep spirits away that caused nightmares.

Garden notes: To many ranchers of semi-desert areas, rabbitbrush is considered a weedy species and large tracts of sagebrush and rabbitbrush are "chained over" to create livestock grazing. Grazing is what created dense stands of rabbitbrush and sagebrush as the livestock avoids eating these, leaving them to grow into thick stands. Rabbitbrush is valuable for xeric landscapes of the hottest, ugliest soil sites. Their beautiful sunny flowers in late fall also provide valuable food to some species of *Lepidoptera* (moths and butterflies).

Asteraceae Family — Sagebrushes — *Artemisia* spp.

Mountain sagebrush flower (top left) and three-toothed leaves (right)

Silver sagebrush

Silver sagebrush leaf (oval)

Black sagebrush leaf

Sunflower Family

MOUNTAIN SAGEBRUSH

Artemisia tridentata ssp. *vaseyana*
(*Seriphidium vaseyanum*)

SILVER SAGEBRUSH

Artemisia cana ssp. *cana*
(*Seriphidium canum*)

BLACK SAGEBRUSH

Artemisia nova (*Seriphidium novum*)

Description: **Mountain sagebrush** is a grayish green wintergreen shrub up to 3 feet tall. Leaves are long and broadly linear with three-toothed tips. Flowers are feathery plumes of small brownish gray clusters in late summer. Found in dry, but richer soil parks of the montane and subalpine. **Silver sagebrush** is similar in appearance to mountain sagebrush, but leaves are longer and skinnier, rarely 3-lobed. Found in moister areas than mountain sagebrush, but found in hybrids with mountain sagebrush on drier grounds (have long skinny leaves with 3-lobed tips). **Black sagebrush** is smaller, generally up to one foot tall, with a black appearance (less hairy than other species), growing in large masses on drier stonier ground at lower elevations in montane.

Medicinal uses: Though having a similar smell, sagebrushes are not related to the garden Mint Family sages (*Salvia* species). Most of the members of this species have been used as medicine for its bacteriostatic, diaphoretic, and bitter properties. It can be used as a topical skin wash and insect repellent. Native peoples used a tea for colds, coughs, digestive complaints and to stop internal bleeding. Today, concerns about toxicity have diminished its internal use.

Edible uses: The small bitter seeds can be roasted, eaten raw or dried and ground into a meal and cooked in soups and stews. Crushed leaves were mixed with meats to maintain good odor. Its volatile oils added fragrance and flavor to liquors. Dried leaves were added to smoking mixtures.

Notes: Native Americans use several species as a ceremonial smudge, to physically and spiritually cleanse the body and environment of impurities and evil spirits. The aromatic volatile oils were used in hair tonics and shampoos, and as moth and flea repellents. Fresh leaves were packed into footwear as a foot deodorant, and the shaggy bark used in the winter as insulation. A green dye was produced from the whole plant set with an alum mordant. Sagebrush was often used in sweat lodges. The floor was covered with fresh branches, fresh or dried leaves placed on the hot stones if someone was not feeling well, and sometimes fresh leaves were stuffed in the nostrils. Sagebrush was used in fire making. A "slow match" was a twisted or braided piece of sagebrush bark carried while traveling. The Uintah Utes twisted sagebrush bark to create skirts, poncho-type shirts and leggings (for the winter), and sandals (with softened bark placed inside for even more warmth).

Garden notes: The woody sagebrushes are valuable shrubs for erosion control of dry, not so steep sites. The big sagebrush (*A. tridentata*) does not do well in our higher elevations, so try and get the **mountain sagebrush** or **silver sagebrush** although they are not easy to find in nurseries.

Caution: Some species cause allergic reactions and dermatitis. Some are considered toxic.

37

Berberidaceae Family   Oregon grape   *Mahonia repens*

Oregon grape habit and flower (inset)

Oregon grape fruit

Barberry Family

Oregon Grape
Mahonia repens (Berberis repens)

Other Names: Holly grape, creeping Oregon graperoot, creeping barberry, creeping mahonia, mountain holly, Yerba de Sangre

Description: Low, wintergreen, sometimes spreading woody plants up to 18" tall. Leaves pinnately divided into spiny-edged leaflets (like holly), turning brilliant red in fall and persisting through winter. Flowers yellow in elongated clusters, highly aromatic (think honey), sometimes hiding amongst the leaves. Berries blue with whitish bloom, in clusters like grapes. Found in open areas and forests from montane to lower subalpine. Blooms May to June.

Medicinal uses: The roots and yellow main stems contain berberine alkaloids that act as a bitter stimulant, antiseptic and antibacterial. It was used by Native peoples as a blood purifier, general tonic and remedy for a variety of illnesses. Today, it is used in blood purifying formulas for cancers, tumors, arthritis and skin conditions. The tea or tincture can stimulate liver function and bile production and is used to treat poor fat digestion, constipation, diarrhea, dysentery, food poisoning and a wide variety of other liver and digestive disorders as well as respiratory ailments such as pneumonia and bronchitis. The roots used both internally and externally aid in fighting infections, especially bacterial including *Staphylococcus aureus*. I consider Oregon grape as our local substitute for the endangered Goldenseal and a powerful plant ally.

Edible uses: Fruit production is really sporadic with Oregon grape. Some years plants are loaded, some years the berries are super sweet. The berry is strong and sour, but when mixed with a little sugar or honey has a rich tangy flavor. They are sweetest after a good frost, and Oregon grape "raisins" in late fall can be really tasty. They are high in Vitamin C. Use the berries in jelly and jam, wine, lemonade, sauces (especially with meats). A refreshing drink comes from mashed berries, sugar and water. Try the berries in with soups for a richer flavor.

Notes: Because of its current popularity this plant is now on the "To Watch" list for plants being over harvested. Though locally abundant, be considerate, use restraint and only harvest what you will actually use. A bright yellow dye was made from roots, and stems were used for weaving and baskets.

Garden notes: This is a valuable landscape plant for dry, part shade sites. Use it for erosion control on steeper sites as it has extensive root systems. It can be, unfortunately, somewhat slow growing in stressful situations. It can take nasty soil, and will grow into quite a clump if allowed to. This is a great addition for wildlife, especially the pollinators.

> **Caution:** Pregnant women should not use this plant because it may stimulate the uterus. Long term use is discouraged. It is best not to use this plant in combination with prescription drugs because it may alter the liver metabolism of the drugs.

> Love is like a beautiful flower which I may not touch, but whose fragrance makes the garden a place of delight just the same.
> – Helen Keller

Cornaceae Family | Red-osier dogwood | *Cornus sericea* spp. *sericea*

Dogwood Family

Red-Osier Dogwood

Cornus sericea ssp. *sericea* (*Swida sericea, Cornus stolonifera*)

Other Names: Red willow

Description: Deciduous shrub up to 8 feet tall with opposite branches with bright red bark, sometimes purplish or green. Leaves opposite, egg to lance-shaped; white flat-topped flower clusters appear at branch tips; fruits are berry-like juicy drupes. Found in moist wooded to open sites; very common along river ways of the valleys. Blooms from June to July.

Medicinal uses: The astringent and tonic inner bark was used by Native peoples internally to treat digestive disorders, diarrhea, fevers, colds, headaches, congestion, pain, general weakness and externally for skin problems, poison-ivy rash, eye infections, ulcers, and snow-blindness. The bark has purgative properties that are reduced with drying. The bark was also smoked to treat lung problems. A poultice of the soaked inner bark is said to decrease pain and bark shavings were applied to wounds to stop bleeding.

Edible uses: If you were to try the bitter fruit, you would understand why they are not generally used today. The Native Americans did collect the fruit, but these were usually mixed with sweeter berries or sugar to create a "sweet and sour."

Notes: This is an important ceremonial plant to indigenous peoples. The inner bark is dried and used in smoking mixtures. The flexible branches have been used in basket-making, especially baby baskets. A red dye can be made from the bark. The bark also provides fiber and cordage and when powdered can be used to protect the gums and keep teeth white. The seeds in oil burn well and can be used for lighting. The wood is very strong and was used to make digging sticks, pipe stems, arrows and pegs.

Garden notes: Dogwoods make great landscape additions because of their colorful red stems and white berries in winter, beautiful white flowers in the spring and moderate drought tolerance. They do require a rich soil. Birds will flock to the berries in late summer; they are a good bee plant. Just be aware that to elk and deer in our area they are considered the "ice cream" bush. Natural pruning!

Caution: Several resources say consumption of large quantities of all parts can be toxic; however, we believe the toxicity is mainly in the berries causing nausea and diarrhea.

Indigenous Management

The landscape that greeted white settlers was not necessarily as "wild" as first thought. Indigenous peoples "managed" many native plants to maximize health and yields. They used a variety of techniques including burning, irrigation, weeding, tillage and pruning. Fire was used to manage many basketry and food producing plants and maintain desired habitat for game. Many tribes saved "the best" seeds for propagation and actively cultivated some native plants. As a whole the Rockies and Great Basin were not as intensively managed as coastal California, the prairies, or the agricultural areas of the Southwest and East. The book *Tending the Wild* by M. Kat Anderson is a great resource for information on this topic.

Buffaloberry

Elaeagnaceae Family — *Shepherdia canadensis*

Buffaloberry fruit, leaf and twig

Buffaloberry young, leafy growth

Buffaloberry fruit and characteristic rusty spotting on leaf back

Oleaster Family

BUFFALOBERRY

Shepherdia canadensis

Other names: Canada buffaloberry, soopolallie, soapberry, russet buffaloberry

Description: Spreading deciduous shrub up to 8 feet tall, thornless, with twigs and undersides of leaves covered by brown scales. Leaves opposite, elliptical, leathery and intensely green above, brownish below. Shrubs dioecious (plants male or female) with tiny chartreuse male or female flowers in clusters at the leaf axils appearing early spring before the leaves. Female plants bear translucent red to yellow berries. Found in open lodgepole pine forests and openings from montane to subalpine. Blooms May and June.

Medicinal uses: Native people had many uses for buffaloberry. The plant decoction was used externally as a wash or poultice for sores, cuts, swellings, broken bones, aching limbs and arthritic joints. A bark infusion was used externally for sore eyes. The stems decocted were used as a stomach tonic and to treat constipation, high blood pressure and venereal disease. The roots were used to treat constipation, hemorrhages, as an aid in childbirth and in the treatment of TB and coughing up of blood. The berries were eaten as a treatment for high blood pressure, stomach cancer and to ease labor pains. Berry juice was ingested for digestive problems and constipation and applied externally for acne and boils. The plant is little used today.

Edible uses: Yes, these are bitter berries, but will sweeten after a couple of frosts. Cooking often helps sweeten them as well. I find that my taste buds go through a series of sweet, rich, bitter when I eat them. Native Americans used the berries extensively because they were locally abundant, using them raw, boiled, formed into cakes and fried over a fire to be used in the future. Because of the saponins in the berries, they foam when beaten. I love Indian ice cream (see accompanying recipe). The berries can be made into a refreshing drink if plenty of sugar is added.

Notes: The crushed fruit can be used as a soap substitute. The branches decocted were used as a hair tonic for curling and dyeing hair.

Garden notes: Buffaloberry, like legumes, can fix nitrogen via a symbiotic relationship its roots form with certain soil bacteria forming nodules on the roots. Some of this nitrogen is utilized by the growing plant but some can also be used by other plants growing nearby. Established plants are drought resistant and make a nice understory in dappled shade areas

Caution: These berries have a low concentration of saponins. Some saponins (soapy substances) can be irritating to the digestive system and large amounts, especially of the berries, should not be consumed. Cooking breaks them down.

Indian Ice Cream

½ cup buffaloberry berries
½ cup water
2 tablespoons sugar

Mix berries and water in a grease-free, non-plastic bowl. Beat mixture until it is the consistency of beaten egg whites, add the sugar as it starts to foam. If the berries are ripe, the foam will be pink. However, even with sugar added, it will be sour and bitter. This is an acquired taste – it is so bitter you should test it before serving. To make it tastier, flavor with cinnamon and sugar, or try serving it over iced chocolate cake.

When picking buffaloberry berries, don't pick into a plastic bag or any other container that has a residue of grease or oil. Like egg whites, the mixture won't get stiff if any oils are present.

—*"Wild Berries of the West"*

Ericaceae Family Kinnikinnick *Arctostaphylos uva-ursi*

Kinnikinnick leaves, flowers and last season's fruit on same plant in early spring

Ripe kinnikinnick fruit

Kinnikinnick habit

Heath Family

KINNIKINNICK
Arctostaphylos uva-ursi

Other names: Bearberry, uva ursi, common bearberry

Description: A trailing, evergreen shrub up to 6 inches tall, sprawling up to several feet while rooting along the reddish stems. Alternate leaves are spoon-shaped, rounded at the tip, dark green and glossy. Flowers are white to pink bells in small clusters producing bright red berries. Fairly uncommon in our area. Found in open, dry, rocky coniferous forests of montane to lower subalpine. Blooms May to June.

Medicinal uses: The leaves of this plant have been used for centuries by Native peoples around the world. It is still used by herbalists today to treat chronic and acute kidney infections and kidney and bladder stones. The leaves, fresh or dried, in tea are disinfecting to the urinary tract and most helpful for alkaline urinary tract infections used with an alkaline diet (vegetable based). A strong leaf tea can be used in a sitz bath after childbirth to reduce inflammation and prevent infection and as a wash for skin irritations, rashes, thrush, burns, an eye wash, a mouth wash for cankers and sore gums and poultice for back and rheumatic pain. A leaf tea can be vasoconstricting to the uterus and may be helpful for painful and heavy menstruation. This is strong medicine with several cautions for use, so do your research or work with an experienced practitioner.

Edible uses: I remember the first time I tried a Kinnikinnick berry and I thought "Hmm, that is pretty bland". Besides the fact that the plants are uncommon and the berries nothing special, I would leave them for the wildlife to enjoy. I read that when cooked slowly the berries pop like popcorn.

Notes: The leaves and fruit provide brown dyes. The leaves were used to tan hides. The dried fruits were used in rattles and as beads. The mashed berries were used to waterproof baskets. The name Kinnikinnick actually comes from its frequent use alone or in mixtures as a smoking herb. The dried leaves are used in smoking mixtures as a substitute for tobacco, to treat headaches and ceremonially for its mild narcotic effects that are said to help calm and clear the mind to receive visions and guidance.

Garden notes: Although not common in our area, Kinnikinnick certainly makes an excellent addition to our landscapes as a groundcover for hillsides and shady habitats. It tolerates dry and poor soils. Because of its mat-forming habit Kinnikinnick helps control erosion.

> **Caution:** Not to be used during pregnancy, by children and those with kidney disease. Large doses may lead to nausea and vomiting. Long term use is discouraged and can cause stomach irritation. Treatment should be short lived (less than seven days) and used with an alkaline diet.

Flower Thieves

There are "nice" pollinators and nectar gatherers and there are "naughty" ones. The flowers of Arctostaphylos have unique hollow anthers with an opening on one end and pollen inside the tube. The most efficient way to get the pollen is by sonification or buzz pollination. Bumblebees and many solitary bees grab onto the flower and rapidly beat their wings, "buzzing" the pollen out of the anther. These are the "nice" pollinators. The "naughty" flower thieves may be beetles or other bees, wasps or sometimes even flies that bite the base of the flower to gain entry to the pollen or nectar. They never transfer pollen from flower to flower.

Ericaceae Family — Huckleberries — *Vaccinium* spp.

Broom huckleberry habit

Dwarf bilberry habit and flower

Broom huckleberry's red fruit

Low bilberry's blue-black fruit

Heath Family

Huckleberries
Vaccinium spp.

Other names: Bilberry, blueberry, whortleberry, grouseberry, grouse whortleberry

Description: Our local huckleberries are low shrubs, rarely over a foot tall with deciduous alternate leaves with small marginal teeth; urn-shaped flowers are white or pink typically born singly along the leaf axil and hiding under the foliage; berries blue, black or red, all with a flat end. Here are a few common locals:

Broom huckleberry (*Vaccinium scoparium*) has strongly angled green branches, often broom-like habit, leaves widest at or below the middle; berries are bright red, sometimes slightly purplish. Often found in dense conifer sites up to alpine.

Low bilberry (Mountain blueberry) (*V. myrtillus*) also has greenish angled branches, but coarser and not broom-like; leaves broadest at or below the middle; berries are blue-black. Found in montane and subalpine forests under spruces.

Dwarf bilberry (*V. cespitosum*) is a spreading shrub with yellowish to reddish, slightly angled branches, lance-shaped leaves are widest above the middle; fruit large blue berries with whitish bloom. Found in moist to dry wooded to open sites from montane to alpine.

Medicinal uses: The leaf tea is useful in lowering and stabilizing blood sugar levels in diabetes and hypoglycemia and can also be used to treat nausea and increase appetite. The leaves are similar in their chemistry to Kinnikinnick, only less astringent, so they are antiseptic, astringent, diuretic and tonic for treating urinary tract problems and can be an alternative to cranberries in treating urinary tract infection. They are also said to help prevent uric acid formation in gout. The berries are high in vitamin C, antioxidant flavonoids and anthocyanosides (the blue pigments). Anthocyanosides have been shown to strengthen capillaries and are helpful in varicose veins, hemorrhoids, enhancing poor circulation and improving night blindness. Low bilberry is the most widely used and studied but Michael Moore says all species are interchangeable so use what grows near you and enjoy!

Edible uses: Basically, anything you use blueberries for, you can substitute these amazingly rich and flavorful wild cousins. This is definitely where the saying "big things come in little packages." One little berry will FILL your mouth with flavor. You will definitely have to work to gather these little gems. There are huckleberry pickers; a container with something like a comb on the end that you swipe through the bushes. Seems to be a little impersonal for me, but it is quick. Here is one secret to successful huckleberry harvesting: gently lift the plant up to find all the berries hiding underneath (especially for broom huckleberry). The leaves of all make a tasty tea; my favorite is low bilberry. Berries make good wine.

Notes: The fruits are loved by wildlife and the plants provide food for a number of insects. The berries produce a navy blue dye for fabrics and ink.

Caution: The leaves contain high amounts of tannins, so they should not be used for long periods. Avoid in pregnancy or if on anticoagulant medications.

Fagaceae Family — Gambel oak — *Quercus gambelii*

Spring Gambel oak habit

Gambel oak fruit

Oak Family

Gamble Oak
Quercus gambelii

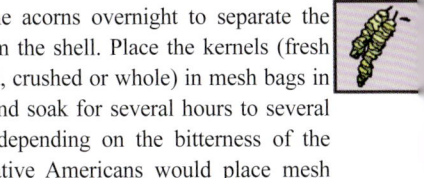

Other names: Scrub oak, Gambel's oak

Description: A deciduous tree or shrub up to 20 feet tall with deeply lobed 2-4" long alternate leaves and small acorns on which the cap covers 1/3 or more of the nut. Flowers are tiny, without petals, and either male or female, but with both sexes on the same plant. Males are the conspicuous flower in Gambel oak; dangling green clusters in early summer. Found on dry slopes in montane.

Medicinal uses: The high tannin content of oak, especially the galls, makes this a powerfully astringent plant that also contains chemicals with actions similar to aspirin. A tea of the bark or leaves and twigs is used internally for diarrhea, bleeding hemorrhoids, dysentery, gum inflammations and sore throats and externally for first and second degree burns, inflammations, poison ivy, abrasions and cuts. It has a clotting, shrinking and antiseptic effect. The quercin in oak can strengthen capillaries and its tannic acid is anti-viral and anti-bacterial. This is a good back country first aid plant to know, using the leaves as a poultice or wash for insect bites and wounds or chewing a piece of the bark for toothaches or making a leaf or bark tea for diarrhea. Because of the strength of its constituents, I recommend using it only for specific conditions and not for long-term internal use.

Edible uses: Gambel Oak is considered a "sweet" acorn (most of the "white" acorns are sweeter than other species), but they are still too bitter to eat raw. Collect acorns in the fall when they have fallen from the trees (Hall suggested raiding squirrel caches), but don't wait too long or worms will invade the nuts. Shell them at once, do not store them with the shell on or worms are almost guaranteed.

Soak the acorns overnight to separate the nut from the shell. Place the kernels (fresh or dried, crushed or whole) in mesh bags in water and soak for several hours to several weeks depending on the bitterness of the nut. Native Americans would place mesh bags of crushed nuts in fast running streams. Sometimes adding wood ash can speed up the process, or kernels can be boiled in several changes of water until the water is no longer brown. Harrington boiled the kernels whole or broken for 45 minutes, changing the water several times. The leaching is best as it removes the tannins but leaves the nutrients. Some tribes added a special clay to remove bitterness; powdered gelatin may also work. After leaching, the kernels can be dried by roasting. The roasted kernels taste a lot like other nuts with a pleasant sweet flavor. Eat them as snacks, bake in cookies and cakes, or dip them in syrup and eat like candy. They can also be brewed for a coffee-like drink. The dried kernels will keep several months. The kernels can be ground into flour and used in breads. The flour is best when mixed with cornmeal or wheat flour. A mix of ½ acorn and ½ whole wheat make great crunchy pancakes. Acorn bread was important in Native American diets. It was a brick red color from the addition of local clays in the proportions of 1 clay to 20 acorn flour to make a stiff dough. Chemists say the iron in the clay helps offset the high protein content in the acorns. The dough was then wrapped in fern leaves and placed in ashes. After baking with a slow heat, the unwrapped bread was black with fern prints on it. It is really sweet and tasty.

Garden notes: I love the fact that this rugged and uniquely structural tree is finally finding its way into the nursery trade. You will also have success with acorns; just place a handful into the area you want them to grow and several will sprout. Weed out the ones you do not want. Be warned that these are vigorous growers, creating widely

Fagaceae Family — Gambel oak — *Quercus gambelii*

Gambel oak leaf and male flower

 Oak Family

Gamble Oak
Quercus gambelii

Garden notes (continued):

spreading clones that can be next to impossible to eradicate if you decide you do not want them. Their advantages far outweigh their disadvantages by providing wonderful habitat for wildlife, being drought tolerant and providing food sources for several insects including the Colorado hairstreak butterfly. Give yourself ample time planning where to place these beautiful trees.

Notes: Flowers are used in flower essence. Oak wood was traditionally used to make ax and hoe handles, digging sticks, weaving tools, bows and arrows, baby cradles, and ceremonial bullroarers.

Caution: The tannic acid can be potentially poisonous with long term use. Cattle and sheep have died from eating leaves or large amounts of acorns.

Gambel Oak Bumps and Lumps

On your next excursion into Gambel oak habitat, take note of how these trees are often growing in clumps. Young stands appear as "dog-hair" stands of skinny straight trunks. Older stands will be one or two majestic trees with younger stems popping up from the outer edges. These stands are clones and are growing from a unique structure called a lignotuber as well as interconnected rhizomes. The lignotuber is a woody mass of tissue (if you scrape away duff from the base of the trees you can often see this swollen mass of growth toward the base of the tree trunk) where adventitious buds grow and form new stems. Lignotubers are found on many fire-adapted species. The top of the plant can be destroyed by fire but the underground lignotuber will quickly re-sprout as well as some re-sprouting from rhizomes.

While you are wandering through the oak, check out the leaves and stems for bumps and lumps. Gambel oak are hosts for gall wasps (specifically cynipid wasps) using oak tissues as shelter for their developing larvae. It is still a mystery as to how these tiny insects create these artistic and bizarre growths on leaves and stems, but most scientists lean toward the chemicals in insect saliva that basically hijack plant tissues to form abnormal structures. These typically do not hurt the oaks.

Currants and Gooseberries

Grossulariaceae Family — *Ribes* spp.

Golden currant habit

Golden currant flowers smell like cloves

Golden currant fruit and "pigtails" on fruit

Golden currant leaf and flower

Currant/Gooseberry Family

CURRANTS AND GOOSEBERRIES
Ribes spp.

Description: All the *Ribes* spp. are shrubs up to 8 feet tall, with or without spines or bristles along or at nodes along the stem; alternate, maple-like leaves; flowers are tubular to saucer-shaped, flared and white to pink to reddish-orange to maroon with 5 petals and 5 sepals All produce edible fruits, varying in delectability, and they all have an attached "pigtail" (dried remains of the flower). We will break out a couple of our most common ones.

Golden currant
Ribes aureum

Other names: Black currant, clove bush, buffalo currant, Missouri currant, yellow currant, wax currant, squaw currant

Description: Bushes 5 to 8 feet tall; stems lacking thorns; leaves smooth and shiny, three-lobed; flowers bright yellow and highly fragrant (cloves!), long and tubular; berries black or red. Found on moister sites in lower valleys.

Medicinal uses: Not only is this the best tasting of these berries, it also has some history of medicinal use. The inner bark is considered anti-inflammatory and astringent. Dried and pulverized, it was made into a poultice or sprinkled on sores and snakebites. An inner bark decoction was used to treat leg swellings. The root, crushed and boiled or a piece chewed was used to treat sore throats. A berry jelly is healing for burns and soothing to sore throats. The leaves, bark and root of all *Ribes* are largely diaphoretic, astringent and diuretic. Prickly black currant (*Ribes lacustre*) is the most medicinal and most commonly used for medicine of the currants. This species is not common in our area.

Edible uses: Because of its larger, flavorful, juicy berry and location next to water this berry is a favorite for many. Lewis and Clark expressed a liking for this berry. The dark berries can be eaten raw, made into delicious jams and jellies, pies and sauces. I have even added them to ice cream. Yummy! The berry juice can be mixed with apple juice for a nice drink. Harrington talks of such an abundance of nectar on the flowering bushes that when it is shaken at the right stage, nectar drops rain down. Because the flowers smell so heavenly, I thought they would taste just as good. They are not clove flavored, but are nice and sweet.

Garden notes: The first time I saw this bush was along the fence of an old homestead. After hesitantly trying the berries I vowed to find more of them! The flowers absolutely put me in heaven with their scent. If you have a fairly moist site, or even a good shady site that will keep the shrubs protected from the sun, these plants transplant well from the wild. They are also very common in nurseries. As an extra bonus, you will revel in the blazing red fall foliage.

Red prickly currant
Ribes montigenum

Other names: Mountain currant, gooseberry currant, mountain gooseberry

Description: Sprawling shrub up to three feet; spines three at a node; leaves palmately lobed and somewhat sticky, alternate; salmon pink flowers saucer shaped; berries bright red. Found in dry forests and rich meadows of the subalpine.

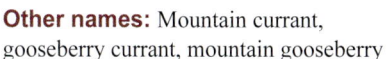

Medicinal uses: I can find no record of this species being use medicinally.

Edible uses: In a good year on Rabbit Ears Pass, where these seem to be prolific, I can fill my bags with plump, rather tart, but tasty berries. Other years there is nary a berry to be found, or those that appear

Currants and Gooseberries

Grossulariaceae Family · *Ribes* spp.

Red prickly currant flower

Red prickly currant fruit and prickly stem

Red prickly currant habit

Wax currant habit

Wax currant fruit (above (oval), leaf and flower (below oval)

Currant/Gooseberry Family

Currants and Gooseberries

Ribes spp.

Red prickly currant

Ribes montigenum

Description (continued):

are so tart and pulpy that they are unusable. Other authors have said how much they like the wax currant (*R. cereum*). I find this one to be very bitter, pulpy and dry. Maybe it is where I usually find them (dry, rocky areas).

Caution: Wax or Golden currant has been noted in several books as being emetic if too many are eaten. The Native Americans said they got the name bear currants because Bear did not like the people eating his berries, so he put this taste in them so they wouldn't like them.

Try them all; no currants are poisonous. Just look for the little "pigtail" at the end of the berry. Use all currants like you would store-bought fruits. Also try the nectar rich flowers of *R. montigenum*.

Reading the Land Through the Plants

When I read something Joseph Barrell wrote in *Flora of the Gunnison Basin*, the Currant Family clicked for me. He talked about how the different species are markedly different both ecologically and geographically, as well as morphologically. I scanned my field notes and realized that *R. cereum* was indeed the currant of the hot, dry rocky places. *R. montigenum* was the "mountain" currant, growing in the meadows of Rabbit Ears Pass. And *R. wolfii* was that undershrub I kept finding in the shade of coniferous forests. This discovery only solidified my belief that through understanding more about our plant friends we can see the landscape with new eyes that are more in tune with and in sync with the land.

Drying Berries

Wash the berries and place in a colander or cheesecloth bag. Dip berries in boiling water for one minute. Blanching promotes even drying. After blanching, spread the berries on paper towels to dry, then place in single layer on screens covered with cheesecloth and cover with another layer of cheesecloth. **Sun drying**: Dry outside in an airy place, check them often and turn them. They should dry in about two days. When finished they should be leathery. Bring the berries in at night so they don't collect moisture. **Oven drying**: Use the lowest temperature possible. Maximum temperature should be 140°F. Prop the oven door open a half inch or more to allow moisture to escape. Check frequently and expect them to become leathery in about eight hours or less. **Fruit leather**: Puree fruit in a blender, sweeten if desired. Line a cookie sheet with parchment paper and spread the puree evenly 1/4 inch thick over sheet. Follow "sun drying" or "oven drying" techniques above to dry leather. If leather bottom is not drying properly you can peel it from the paper and flip once. Times will vary according to the thickness of the leather, but will be similar to times noted above. Roll up leather, and store in cool area or keep in freezer for longer term storage.

Rhamnaceae Family — Elkbrush — *Ceanothus velutinus*

Elkbrush habit

Elkbrush flower

Buckthorn Family

ELKBRUSH
Ceanothus velutinus

Other names: Snowbrush, velvety buckbrush, sweet bush, deerbush, tobacco brush, mountain balm, sheep-herder tea, sticky laurel, redroot, New Jersey tea, mountain laurel.

Description: A sprawling shrub up to 6 feet tall with aromatic, shiny, sticky, ovate leaves with fine serrations that roll downward (especially when it's very dry). Look closely at the leaves for the 3 predominate veins from the leaf base to the outer margins of the leaf. Flowers are pyramid shaped clusters of sweet smelling cream flowers. Large areas of elkbrush can be found on drier and open slopes of the montane.

Medicinal uses: The leaf tea was used for treating inflamed tonsils, enlarged lymph nodes, non-fibrous cysts, and enlarged spleens. Native Americans also chewed the leaves to relieve inflammations and irritations of the mouth and throat. The leaves contain saponins and were used as a skin wash for treating various skin problems. The root is high in tannins, making it a good astringent to membranes. It also has expectorant and antispasmodic actions. It is a good gargle and mouthwash for sore throats and mouth sores and a decoction for bronchitis, asthma and coughs. The most popular use of the root today is as a lymphatic stimulant and tonic, being used to treat longstanding problems like ovarian cysts, fibroid cysts, tonsillitis, enlarged lymph nodes and anywhere impaired lymph flow is suspected. The roots are extremely hard when dry so they need to be processed immediately in the field.

Edible uses: The leaves make a good substitute for commercial black tea and are actually a little sweeter. The flowers also make a pleasant tea.

Notes: The name red root comes from the blood red color of the inner root bark. All parts of the plant contain saponins and make a nice soapy lather when crushed and rubbed in warm water, leaving the skin soft and fragrant. The leaves have been used as a tobacco substitute. The seeds may lie dormant for 200 years until fire activates the germination process.

Caution: Not to be taken by those with blood disorders or allergies to aspirin. Use in moderation is recommended during pregnancy.

> To rest, go to the woods
> Where what is made is made
> Without your thought or work.
> Sit down; begin the wait
> For small trees to grow big,
> Feeding on earth and light.
> Their good result is song
> The wind must bring, the trees
> Must wait to sing, and sing
> Longer than you can wait.
> Soon you must go. The trees,
> Your seniors, standing thus
> Acknowledged in your eyes,
> Stand as your praise and prayer.
> Your rest is in this praise
> Of what you cannot be
> And what you cannot do.
>
> Wendell Berry
> "The Farm" IX
> *A Timbered Choir:*
> *The Sabbath Poems*
> *1979-1997*

Serviceberry

Rosaceae Family — *Amelanchier alnifolia*

Serviceberry leaf and flower

Serviceberry fruit with "star" on end

Serviceberry shaded habit (above) and open sunny habit (below)

Caution: The leaves and pits contain poisonous cyanide-like compounds which are destroyed by cooking or drying.

Rose Family

SERVICEBERRY

Amelanchier alnifolia

Other names: Saskatoon, June-berry, shadbush, sarvis berry, shadblow

Description: A fairly dense shrub up to 15 feet tall. Leaves alternate, ovoid, with large serrations toward tips of the leaves, the leaf edges fairly smooth toward the petiole. Flowers in clusters, five white petals that are rather long and unkempt, many stamens, bark smooth and gray. Fruits are purple to black with a whitish bloom, with small seeds and a "star" on the end. Found in open woods and hillsides from foothills to montane, often found in dense shrublands with chokecherry and hawthorn. Blooms May to June.

Medicinal uses: Native peoples used this plant for a variety of complaints from treating snow blindness to colds and fevers. It was combined with other plants in contraceptive recipes. The fruit juice decoction was used for upset stomachs and restoring children's appetites and externally as ear drops. The bark and stems can increase sweating to reduce fevers, treat colds and lung infections, expel afterbirth, regulate menstruation and be a laxative.

Edible uses: The tasty berry was probably the most important berry crop for the Native Americans because of its versatility, nutritional value in winter months and ease of gathering. The fruit is high in iron and copper. They taste and can be used just like blueberries!

This drying process described by Willard produces berries that can last in storage several years: into the bottom of a large spruce tub, place a layer of berries, red hot stones on top of them, then another layer of berries, then hot stones, layering like this until the tub is full. Leave for 6 hours or until completely cooked. Crush between hands, spread on thin planks of wood tied together, and place on a slow fire.

Collect the juice for dyes, or make a tasty juice by adding a little sugar. Leave to dry for 2-3 days. Today we simply dry them using a dehydrator or placing them in the sun and dry like raisins. I love eating the "servins" (serviceberry raisins) in the fall after they are a nice little dried pouch of blueberry-flavored sugar. Yummy!

Notes: The wood is very hard and was used for arrows, pipe stems and snowshoe frames, with smaller branches used for basket rims and tipi stakes. Serviceberry was important in many Native American ceremonies; the Sun Dance was always held when the berries ripen.

Garden notes: These shrubs are excellent additions to landscapes, providing food and shelter to birds, are attractive structural forms, can take our nasty clays and are very drought tolerant. Transplanting from the wild works well, but expect a slow recovery time.

Spiced Pemmican

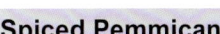

4 cups beef jerky, broken into small pieces
4 cups serviceberries
3 Tbs butter
3 Tbs brown or white sugar
1/4 tsp powdered ginger
1/4 tsp cinnamon
1/4 tsp cloves

Mash serviceberries and mix with sugar, spices and butter. Bring to a boil and simmer for five minutes. When cool, mix with jerky and spread thinly in a baking pan. Dry overnight in oven at 150°F. You can vary the pemmican by adding unsalted nuts, sunflower seeds, or other berries. You may prefer it with more sugar than the recipe calls for.

— *Wild Berries of the West* page 206

Rosaceae Family — Hawthorns — *Crataegus* spp.

Hawthorn habit

River hawthorn habit

Hawthorn fruit

Hawthorn thorns

Hawthorn leaf and flower

Rose Family

Cerro Hawthorn
Crataegus erythropoda

River Hawthorn
Crataegus rivularis

Description: A woody shrub up to 12 feet tall with stout thorns (up to three inches long!) along the branches. Leaves are ovate with wedge-shaped bases and coarse serrations, with the teeth all pointing toward the leaf tip. Flat topped clusters of white, 5-petaled flowers appear in spring, forming red fruits with several small seeds in the fall. The fruits look like miniature apples. Hawthorns are found either on drier slopes (*C. erythropoda*) or along rivers (*C. rivularis*) of the montane.

Medicinal uses: The flowers and berries of many hawthorn species have long been used as a heart tonic and modern research supports this use. I have used our species with good results. They have slow, gentle actions that strengthen weak functions or decrease excessive functions as needed. They are good for many heart and circulatory problems like hypertension, angina, arrhythmias and arteriosclerosis. They are especially indicated in the treatment of weak heart combined with high blood pressure or to treat a heart weakened by age. They are taken internally as a tea or tincture. The berries are high in flavonoids and used to strengthen connective tissue and improve digestion. The astringent bark has been used to treat malaria and other fevers.

Edible uses: Think of a bland, mealy apple and that is what hawthorn fruits are like. They can be used to create delicious candy, and pretty rose colored jams and jellies. Steep the fruits to make a pleasant tea, cold drinks or wine. Because the berries are high in pectin, they can be used to help gel other fruits such as serviceberry and chokecherry. Historically the fruits were used fresh or dried for the winter, or added to pemmican. They were also cooked, the seeds were removed and the flesh pounded into a pulp, then dried into cakes as berry-bread. These could be used in soups or eaten with deer fat or marrow. Quality varies widely from tree to tree. When a sweeter one is found, guard it, and don't let anyone know!

Notes: The tough thorns were used as awls and fish hooks. The wood is very hard and used to make durable walking sticks.

Garden notes: There are many hawthorns available in the landscape trade. They are very hardy, drought tolerant shrubs and small trees and offer valuable fruits for birds in the winter. Although the native species has big thorns (great shrub to put where you want no foot traffic!), there are now thornless cultivars with similar cultural needs. Just remember, many cultivars may not have the same medicinal properties or edibility as the natives.

Caution: The stout thorns are at eye level and can cause blindness. Be careful! Too many fruits can cause nausea, vomiting, and diarrhea. Because hawthorn can affect blood pressure and heart rate, persons with heart disease should consult their medical practitioner before using.

Chokecherry

Rosaceae Family — *Prunus virginiana* var. *demissa*

Chokecherry habit

Chokecherry fruit

Chokecherry leaf and flower

Rose Family

CHOKECHERRY

Prunus virginiana var.*demissa (Padus virginiana* ssp. *melanocarpa)*

Other names: Western chokecherry, black chokecherry, chokeberry, capulin

Description: A deciduous shrub up to 15 feet tall producing beautiful elongated clusters of 5-petaled white flowers in early spring. Leaves are ovate with fine serrations and small glands (they look like a small bump) at the base of the petiole. The shiny black berries are in elongated clusters and have a large cherry-like pit. Found in dry to moist open sites up to montane. Often forms dense shrublands with serviceberry and hawthorn. Blooms May to June.

Medicinal uses: This was a popular and versatile remedy of Native Americans. The inner bark of this shrub was used to make syrup or tea for coughs, chest colds and bronchitis. The inner bark was also used to treat headaches, heart problems, fevers and to expel worms. The bark and dried berries are astringent and can be used in the treatment of diarrhea, hemorrhoids and sore throat. Today, the inner bark is also recognized as a sedative and tonic for many kinds of chronic irritated conditions of the heart, lungs and gut.

Edible uses: Because of the abundance of these berries, they were a very important food for early tribes, and are still coveted today for jams, jellies, syrup and wine. The name "choke" cherry comes from the extremely bitter taste and astringent qualities of the younger berries. They become sweeter as they mature, and are the sweetest after a frost. They are also sweeter after they have been dried or cooked. See "Caution" before using any parts of this plant. A pleasant juice can be made with the berries, mixed with sugar and water. Historically the berries were dried or cooked, added to dried meat and fat to make pemmican, or pounded, formed into patties and dried in the sun to make berry cakes for the winter. These patties could then be soaked in water, mixed with flour, sugar and water and made into a sauce. The berries were also dried for the winter.

Notes: Chokecherry wood is hard and does not easily burn, so forked sticks were used to carry hot rocks to sweat lodges, and to place hot rocks in baskets of water to boil the water. The wood was also used for digging sticks, tent pegs, backrests and other household items.

Garden notes: This is a fantastic shrub to introduce into a non-formal landscape as it can be rather lanky and "wild" looking unless heavily pruned. The benefits to wildlife are many; berries for birds and many animals, dense foliage for nesting and honeybee heaven during flowering. This shrub is great for erosion control on drier slopes because of its tendency to sucker and form dense stands.

> **Caution:** All parts of the chokecherry, except the flesh of the fruit, contain hydrocyanic acid. Cooking or drying will destroy the poisonous cyanide. The leaves and twigs are poisonous to animals.

> The flower is the poetry of reproduction.
> It is an example of the eternal seductiveness of life.
>
> — Jean Giraudoux

Rosaceae Family — Wild rose — *Rosa woodsii*

Wild rose habit

Wild rose hips (fruit)

Wild rose flower

64

Rose Family

Wild Rose

Rosa woodsii

Other names: Dog rose, Wood's rose

Description: A woody shrub up to 12 feet tall with lanky, sometimes arching stems with recurved thorns (not densely bristly). 5-petaled fragrant pink flowers with many stamens, producing a red fruit (a hip). Leaves are alternate, pinnately compound with a noticeable stipule. Growing in a variety of habitats from montane to subalpine.

Medicinal Uses: All parts of the rose are astringent and useful externally to decrease redness and swelling of wounds and inflamed sore eyes, and internally to relieve diarrhea, reduce labor pain, decrease muscular pains, for sore throats, heartburn and to stop bleeding. The petals or buds and hips are the most frequently used parts as a tea or wash. The hips are used for the colds, chills, flu-like infections, bowel disorders, vitamin C deficiencies, to aid digestion, for gallstones, colic, edema, as a kidney tonic and diuretic. The petals and buds assist the skin in retaining moisture, cell regeneration, healing scar tissue and protecting sensitive skin, making them popular for cosmetics. The fragrance is valued for headache pain, depression, anxiety, as a sedative and aphrodisiac.

Edible uses: Roses are trail munchies! Eat the fresh petals throughout the summer, and the ripe hips (they're best after a frost) in the fall. Avoid eating the seeds as they have irritating hairs, causing "itchy bum." Take the fresh petals and wrap a stick of butter with them. Leave the butter for a couple of days, then use on cornbread, pancakes and other goodies, or candy the petals for a fragrant addition to desserts. The hips can be dried and steeped in a tea, or ground into powder used as a flavoring. Use fresh hips to make jams, jellies, wines, syrup, and candy. One ripe rose hip can have 500 mg of Vitamin C. The flavor and texture of the hips varies widely. Finding a sweet, tender shrub is a gold mine. Very young leaves are used as a spring green. The young buds can also be placed in salads. A rose colored drink can be made from the roots.

Notes: Use the aromatic petals in potpourri. The inner bark can be smoked like tobacco. The old straight wood was used for arrow shafts. A yellow dye is obtained from the inner bark. The petals are used to make rosewater. The hips are very high in Vitamin C (drying and heating can deplete Vitamin C) as well as Vitamins E, B, K, and beta-carotene.

Garden notes: I will warn you that introducing Wood's rose into your garden means you will always have Wood's rose in your landscape! They are aggressive and you will find them sprouting up all over your yard. That said, if you have the room and want to introduce an intensely valuable plant for you and wildlife into your landscape, go for it. A small portion of the root with a good shoot or two on it will get you started.

> **Caution:** The seeds have irritating hairs and should not be ingested.

> Of all the flowers methinks the Rose is the best.
>
> — William Shakespeare

Rosaceae Family — **Thimbleberry** — *Rubus parviflorus*

Thimbleberry habit

Thimbleberry fruit (above)

Thimbleberry flower and leaf

Rose Family

THIMBLEBERRY

Rubus parviflorus (Rubacer parviflorum)

Other Names: Western thimbleberry, Japanese raspberry, salmonberry

Description: Deciduous shrub up to 4 feet tall lacking bristles or spines. Leaves usually over 4 inches long look like a maple leaf. Large white rose-like flowers, 5-petaled, held in loose clusters above the leaves. Fruit looks like a flattened raspberry. Found in moist, rich, shaded to open habitats from foothills to lower subalpine. Blooms May to July.

Medicinal uses: Like all members of the Rose Family, the leaves have astringent properties which make them useful for treating diarrhea, stomachache, vomiting, acne, anemia, strengthening the blood, regulating prolonged menstruation and as an appetite stimulant. Externally, a leaf poultice or wash is used for wounds, burns and to lessen scar formation. A tea of the root has been used in the treatment of stomachic disorders, diarrhea and dysentery, pimples and blackheads and by thin people to gain weight. Like the fruit of raspberry (Rubus species) thimbleberry fruits are strong antioxidants and have value in inhibiting influenza and herpes virus. The fruit, fresh or dried in a tea, can be useful in the early stages of common viral infections.

Edible uses: The first time I read in a book that the berries were "tasteless... coarse and seedy", I thought they were describing a different plant. I LOVE these berries; always sweet and rich and tangy. When fully ripe they can be mushy and difficult to collect. They will usually not be good for drying, unless they are made into fruit leather because they will crush so easily. The berries make excellent jams and jellies. Harrington said the Indians of the Northwest ate the berries with half-dried salmon eggs (hence the name salmonberry). Young shoots can be eaten raw or cooked. Fresh flowers are nice in salads. The leaves, fresh or dried, can be used in teas.

Notes: The large leaves were used as plates, containers, basket liners, and toilet paper. Soap was made from the boiled bark and a purplish blue dye from the fruit. The tasty berries are attractive to many birds and animals and the flowers are valuable to pollinators.

Caution: Use only fresh or fully dried leaves. Wilted leaves can be toxic.

Taste of Plants

The taste of plants can tell us a lot about what actions they might have in the body. Bitter tasting plants (yarrow) stimulate and improve digestion. Astringents (goldenrod) are toning and tightening, helping to stop the flow of things like blood, sweat, diarrhea and help tone tissues. Sweet or bland tastes (rose petals) are usually nutritive, strengthening and tonic. Pungent tastes (wild mint) are stimulating, warming and promote circulation by opening pores. The sour taste (Oregon grape berries) affects liver and gall bladder function and promotes metabolism and circulation. Mineral salty (nettles) affects the kidney, adrenals and urinary bladder by supporting their functions. Though not a taste, plants that are aromatic (mints) are all antimicrobial, decrease inflammation, and reduce fevers.

Rosaceae Family — Raspberry — *Rubus idaeus*

Raspberry fruit and leaf

Raspberry flower

Rose Family

Raspberry

Rubus idaeus (Rubus idaeus subsp. *melanolasius)*

Other names: Red raspberry, wild red raspberry, grayleaf raspberry

Description: Erect to spreading shrubs up to 3 feet tall with biennial branches which are green and bristly in first year, yellow to cinnamon brown and with straight, slender prickles in the second year. Leaves alternate, pinnately compound, flowers white and 5-petaled, sepals reflex backwards. Fruits are red druplets (look just like store bought raspberries). Found in moist to dry open or open-wooded sites from foothills to lower subalpine. Bloom June to July.

Medicinal uses: The leaf and root have long been used for their astringent, anti-inflammatory, antispasmodic and tonic effects, acting most specifically on the reproductive system. The leaf is the easiest part to use medicinally as a tea or tincture and has been traditionally used by expectant mothers during the last part of pregnancy and during labor to strengthen the uterus, ease contraction pain, check bleeding, relieve nausea, and increase milk flow. It is also used to treat irregular and excessive menstruation and is considered an effective woman's tonic. The gentle astringent properties of the leaves are helpful for diarrhea especially in children. The leaves are soothing and tonic to the stomach and bowels and can also help at the first signs of colds and flu. The tea makes a good mouthwash and gargle for mouth inflammations, sore throats, tonsillitis and as a wash for minor wounds, varicose veins, and burns. The berries can be mildly laxative and diuretic, anti-viral and strongly antioxidant. Syrup of the berries is said to be good for the heart.

Edible uses: Raspberries are probably one of the most beloved fruits, along with strawberries. The massive fruits bought in grocery stores don't even touch the intense flavors of the wild fruits. They may be small, but they pack a huge punch! Raspberry fruit clusters drop very easily when they are ripe so I usually use two hands; one under the fruit to catch them if they fall. Use them in any way you would store-bought raspberries. They contain enough pectin to set up jams and jellies. Puree the fruit and boil at a low heat with honey for pancake syrup. Add fresh flowers to salads. Very young shoots are peeled and used raw or cooked. Native Americans ground the fruit, dried this in the sun and stored it for later use. They might mix the dried fruit with cornmeal.

Garden notes: Our wild raspberries are so easy to get established in our landscapes, but growers beware, they are tenacious!! If you have limited room but still want to grow our tasty native, take a shovel and cut the runners off every year to keep them from spreading.

Caution: Wilted leaves can be toxic. Use only fresh or completely dried leaves for tea.

Is It a Thorn, Prickle or Spine?

These all are sharp, stiff and hard structures that deter animals from eating the plant. But what is the difference? A thorn is a stout, sharp, woody outgrowth of the stem or branch. Prickles are short, sharp outgrowths of the epidermis layer, usually more tender than a thorn. The correct term for rose "thorns" is actually rose "prickles." Spines are sharp woody or rigid outgrowths from a stem, leaf, or other plant part.

Mountain ash
Rosaceae Family — *Sorbus scopulina*

Mountain ash habit

Mountain ash flower

Mountain ash leaf and fruit

   Rose Family

Mountain Ash

Sorbus scopulina

Other names: Greene's mountain ash, Western mountain ash, Rocky Mountain ash, Cascade mountain-ash

Description: Deciduous shrub up to 12 feet tall, erect or somewhat spreading, with large clumps (corymbs) of flat-topped white colored flowers in June, followed in the fall by brilliant red or orange-red berries. Leaves are pinnately compound, turning brilliant red in fall. Found in rich, moist areas of aspen or other open forests from montane to subalpine.

Medicinal uses: There is a history of medicinal use for this astringent and diuretic plant among Native peoples and early settlers. A tea of the inner bark or buds was used for rheumatism, arthritis, stomach problems, weak kidneys, diarrhea, nausea, hemorrhoids, and as a blood cleanser and spring tonic. The leaf or berry tea or the chewed bark relieved sore throats and cold symptoms and the berry tea was used for scurvy and worms.

Edible uses: The berries are very bitter when not fully ripe; it is best to let them freeze a couple of times to bring out the sugars Ripe berries can be roasted or dried, and flour can be made from dried berries. Early settlers used the berries in pies, jams, jellies and a bittersweet wine. The berries are very high in Vitamin C.

Notes: Because the berries remain on the shrubs throughout the winter, and usually extend out of the snow, they are valuable food sources for birds and small foragers in the winter.

Garden notes: The native mountain ash is a fairly common shrub in the landscape industry now. It does need a cool, protected area to thrive (note where it grows in its native habitat).

Caution: The fresh fruit contains cyanide related compounds and can be purgative. Drying and cooking neutralizes these chemicals.

Soil Food Web

The soil food web is the community of living organisms in the soil and their relationships to each other and their environment, plants and animals. Bacteria, fungi, arthropods, nematodes, and protozoa along with soil animals like earthworms, millipedes and mites make up this complex ecology. They break down, consume and transform nutrients and shuttle these nutrients around as food for themselves and for other soil life. Healthy soil contains millions of these organisms and the plants depend on them to obtain nutrients and water from the soil, prevent nutrient loss, protect plants from pathogens, and degrade compounds that could inhibit growth. Each type of microorganism plays a unique role in these processes. Our reliance on petrochemical-based pesticides and fertilizers and frequent tillage breaks down the soil food web and decreases the soil and plant health and nutrient value. Adding compost and other organic amendments will improve soil tilth, texture, acration, drainage, and nutritional content. By reducing tillage, we boost the growth of the helpful soil web and suppress many pathogens.

Dr. Elaine R. Ingham, who is considered a leader in soil microbiology research, coined the term soil food web. Dr. Ingham's understanding of the soil food web and how to encourage healthy plant growth and reduce reliance on petro-chemicals is shared at her website, *soilfoodweb.com*.

Salicaceae Family — Quaking aspen — *Populus tremuloides*

Quaking aspen habit

Aspen fall color

Quaking aspen leaves

Quaking aspen catkins

Willow Family

Quaking Aspen
Populus tremuloides

Other names: Aspen, trembling aspen, aspen poplar, quakies

Related species: Cottonwoods (*P. angustifolia*) also contain varying amounts of salicin and populin and can be used in similar ways.

Description: Tall slender deciduous trees up to 60 feet tall; short rounded crown; stunning white bark often marked with black spots and lines. Leaves are alternate, heart-shaped, smoothly serrated, with flattened petioles. Produce catkins of fluffy, cottony seeds. Prefers moist, protected sites from foothills to subalpine.

Medicinal Uses: Quaking aspen has a long history of medicinal use by all peoples. Aspen contain the chemicals populin and salicin, precursors of aspirin, which reduce pain, fever and inflammation. The inner bark is used as teas or tinctures to treat fevers, arthritis, rheumatism, colds, urinary tract infections, diarrhea, indigestion, headaches and to kill parasitic worms. The inner bark is also considered a stimulating tonic for debilitating or weakened conditions. Native people made poultices and washes of the bark and leaves for treating cuts, wounds and rheumatism. They also boiled the bark into syrup which was applied to a broken bone and hardened to serve as a splint. The young buds can be infused in oil as the basis for a healing salve. Syrups from the inner bark can be used as a cough medicine or spring tonic. The inner bark and spring buds contain the highest amount of the active constituents. The spring leaves and soft branches with catkins make a wonderful medicine without the potential damage to the tree that harvesting bark can cause.

Edible uses: All parts of the aspen are bitter except for the sweet sap collected in the spring when the sap is flowing. At this same time the inner bark can be stripped and eaten raw as a sweet treat. At other times of the year, the inner bark is considered a survival food. For Native Americans and their horses it was occasionally their only sustenance during harsh winters. The catkins (the fluffy seeds creating "spring snow" in early summer) and leaf buds are high in Vitamin C, but are also very bitter.

Notes: Leaves and twigs produce gray, gold and brown dyes, and flowers are used for flower essences.

Garden notes: Aspen are one of the most common trees sold in the industry now. Beware they are short-lived trees, prone to many diseases and insects. Providing them optimal growing conditions of adequate moisture and good soil will help maintain (not guarantee) their health.

A BIG Tree!

What is considered the largest living organism on Earth? Aspen!! Aspen form patches (called a genet or clone) of genetically identical trees each (called a ramet), with its own trunk, branches, leaves and a shared root system. It all started from a single seed. This emerging tree sent out a root system with new shoots growing into trees surrounding the original tree. Unless severed, the root system remains connected and assists all the trees in the clone. Trees near a water source can share that resource with other trees. On the negative side, if a tree is attacked by disease or insects it is shared throughout the clone. A remarkable clone in Fishlake National Forest in Utah named Pando (Latin for "I spread") covers 107 acres and contains about 47,000 ramets. Scientists believe it could weigh at least 6,600 tons and be at least 80,000 years old.
Cited: Grant, M. & Mitton, J. (2010) Case Study: The Glorious, Golden, and Gigantic Quaking Aspen. Nature Education Knowledge 3(10):40

Salicaceae Family — Willows — *Salix* spp.

Willow habit

Willow male flower (above), female flower (left)

Caution: Anyone with aspirin sensitivities or allergies should not use willow.

Willow Family

WILLOWS

Salix spp.

Description: There is a wide variety of willow species, many of them crossing to create unique plants. Most willows are shrubs, from a mere few inches (in the alpine) up to 15 feet tall. They are found in moist or seasonally wet conditions from foothills to alpine. All have an unusual bitter aroma when crushed. Leaves are typically lanceolate and serrated. All have catkins.

Medicinal uses: Like aspen, the willow's bark, catkins and young branches contain salicin, populin and tannin. It has been used around the world for centuries to treat inflammation and pain. I like to use the entire young spring branches with leaves and catkins rather than bark in a tincture or dried for a tea. I try to use the plants whose leaves taste most like aspirin. I often gather from several different plants and species to get the highest levels of healing constituents. Just chewing on the branches can be a trail-side first aid remedy. This plant is also antiseptic, astringent, sedative and diaphoretic. Colds, flu, fever, headaches, diarrhea, colic, worms, arthritis, gout, and rheumatism are just a few of the many conditions the willows can help with. This is a plant worth getting to know.

Edible uses: The aspen is in the willow family, so much of what is written for aspen is applicable here. The willows are also considered a survival food. The young shoots, leaves, buds and inner bark are all edible raw or cooked, although very bitter. The inner bark is a little sweeter and can be dried and ground into flour.

Notes: Mother Nature's grocery and hardware store! The Native Americans used the pliable willow branches for pins, pegs, backrests, mattresses, fish traps, fox traps, cradle boards, snowshoes, gambling wheels, walking sticks, stirrups, hide-scrapers, hoops for catching horses, baskets, drums, meat racks, sweat houses, and temporary tipis. The bark was made into twine, used in basketry, finishing nets (very strong when wet). The leaves were soaked in water, and when boiled down produced a green paint. Willow whistles are made in the spring from twisting the bark from the branch and creating holes.

Garden notes: This is one of the most valuable reclamation plants for wetlands. To re-vegetate a moist area, cut willows from the area in early spring. Use younger stems, and cut them about 2 feet long, making sure to keep the terminal part up so that it will be placed in the ground correctly. Simply push them into the ground about 4 inches.

"**Wetland ecosystems** exist in a fragile balance; even subtle changes in the water table can cause irreversible damage. Throughout the Southern Rockies, as in most areas of the West, wetlands are in jeopardy, doomed to increasing fragmentation and dewatering from both climatic changes and a host of human activities. Severe drought can transform a lush oasis for life into a brittle, mud-cracked wasteland in a single season. Much of the human-wrought destruction is the result of transmountain and other water diversion projects, stream channelization, development dredge and fill, peat mining, and degradation related to agricultural activities. Because the region's economic development has been largely concentrated along mountain watercourses and floodplain corridors, ecosystems associated with streams and rivers have been especially hard hit. The future of many mountain wetlands remains uncertain, dependent on our wise stewardship of these important ecosystems."

— Audrey DeLella Benedict *The Naturalist's Guide to the Southern Rockies*

Alliaceae Family — Wild onions — *Allium* spp.

Geyer's onion habit (right) and flower (circle inset below)

Nodding onion flower and habit (below left)

Brandegee's onion (below)

Onion Family

Wild Onions

Allium spp.

Description: There are many species of onions throughout our area, but all have the distinctive onion odor. This is important when it comes to collecting as they can be confused with Death Camas when they are young—a fatal mistake. All onions in our area are perennials with oval bulbs, bell-shaped flowers of 6 tepals clustered atop a leafless stem. Leaves are strap-shaped and strongly aromatic. Habitats vary from extremely arid (*A. brandegeei*) to wet (*A. cernuum, A. geyeri*) conditions, and full sun to part shade from foothills to subalpine. Bloom times vary.

Medicinal uses: Onions are listed as stimulant, carminative, antiseptic, diuretic and expectorant. The juice can be boiled down to make syrups for colds and sore throats. Like its cousin garlic, onions can lower cholesterol. They are anti-bacterial, antiviral and antifungal and can be used in the treatment of cuts, burns, insect bites and stings. The bulbs have also been used to relieve indigestion, gas and vomiting and were said to cure sexual impotency. Crushed, dried bulbs can be used as a snuff for opening the sinuses.

Edible uses: All wild onions can be used just like our store bought onions, and they taste better and are healthier! Some species are stronger than others, so try a few to see which one fits your taste. The bulbs are generally the strongest. If they are too strong raw, cooking will make them sweeter. The leaves can also be thrown in salads, sauces, soups, or nibbled on as trail munchies or placed on your sandwiches. The flowers are generally sweet and can be thrown in salads or on sandwiches, or eaten as trail munchies.

Notes: Yellow, orange and brown dyes can be made from the skins. Flowers are beautiful in dried arrangements.

Garden notes: With the many species of onions in our area, you are sure to find one or two that would fit perfectly in your landscape. I have several species in my garden and have blooms from early spring (*A. brandegeei*) to August (*A. cernuum*). The bees love them and they are excellent companion plants!

Caution: As with cultivated onions, too many can cause poisoning. Eating normal amounts is not considered dangerous. Some people develop a rash by handling onions. **If you are collecting onions, smell every single one. If it does not smell like onions, do not collect it. You could have Death Camas.**

Edible Roots

The Colorado Utes are known to have dug for various roots, using the pointed wooden digging stick. Edible roots include, among others, sego lily, yampah, onions, spring beauty (*Claytonia lanceolata*), and glacier lily. Most were collected in the spring, although they could be dried and stored for later use. They were eaten raw, or if dried, reconstituted for stews or ground as flour for mushes. Digging the tubers with a digging stick aerated the soil, which increased its moisture-holding capacity and prepared the seedbed, likely heightening seed germination rates. Routine burning was also used to increase plant density. Burning and digging together most likely created denser patches of plants, larger numbers and larger gathering tracts.

Cow parsnip

Apiaceae Family

Heracleum sphondylium spp. *montanum*

Cow parsnip large, hairy maple-shaped leaves clasp the stem

Cow parsnip fruit

Cow parsnip habit

Cow parsnip seedhead and flower

Parsley Family

Cow Parsnip

Heracleum sphondylium ssp. *montanum* (*H. lanatum*, *H. maximum*)

Description: A very stout perennial up to 6 feet tall with large (up to 12 inches) hairy maple-shaped leaves clasping the hollow stem. Large flat-topped umbels of cream colored flowers. The whole plant has a rank smell. Found in moist sites from montane to subalpine. Blooms June to August.

Medicinal uses: Many native groups used this plant. A root poultice was used for sores, bruises, swellings and rheumatism. The root or leaf tea was used for a hair tonic, dandruff preventative, sore throats, headaches, epilepsy, colds and flu, nausea and diarrhea, warts, acid indigestion, stomach cramps, TB, epilepsy, asthma, nervous disorders and more. All parts of the plant are antirheumatic, antispasmodic, carminative, febrifuge and stimulant. The fresh seeds or piece of root can be applied to aching teeth. Today, some herbalists also are using dried root extracts to treat nerve damage and paralysis and it is being investigated as a treatment for psoriasis, AIDS and leukemia. This plant has many potential uses that we have yet to rediscover!

Edible uses: Historically these plants were used extensively. The young stems before they flowered were peeled to remove the outer skin and inner pith and eaten raw, steamed or boiled. Very young leaves can be placed in salads, and older leaves can be dried, burned and the ashes used instead of salt. The hollow basal portion of the plant can also be cut into short lengths, dried and also used as a salt substitute. The ripe and unripe seeds can be added to salads (watch out, they are extremely strong). Today the young peeled stems can be frozen, canned or dried, and the roots cooked like parsnips. In my opinion, all parts of cow parsnip have a rank taste to them that just doesn't leave even when cooked. You can be the judge.

Notes: The large leaves can be used as toilet paper. Also take the leaves or a flower infusion and rub on your skin for an insect repellent. Yellow dye is made from the root. Hollow stems can be used for whistles and straws.

Garden notes: Even if you don't want it, cow parsnip will probably find its way to your property if you have the right conditions — rich, moist soils.

Caution: This plant is not to be used internally during pregnancy. Some people are very sensitive to the small hairs found all over this plant and break out in rashes and even blisters. Err on the safe side and wear gloves until you know your sensitivity. The umbel could be confused with other toxic plants in this family. **Make sure you have positively identified this plant.**

Native American Favorites

Native American Ethnobotany rates common cow parsnip as the sixth most widely used species overall, and fernleaf biscuitroot as the number four most widely used medicinal. Yampah and biscuitroot were important Native staple carbohydrate foods. Many of these plants are still very much in use by Native people today, including Porter's lovage and yampah.

Apiaceae Family Biscuitroots *Lomatium* spp.

Wasatch desertparsley flower and leaf

Gray's biscuitroot (left oval)

Fernleaf biscuitroot seed

Fernleaf biscuitroot flower stalk and leaf

Fernleaf biscuitroot habit (below)

Parsley Family

BISCUITROOTS

Lomatium spp.

FERNLEAF BISCUITROOT

Lomatium dissectum

Other names: Fern-leaved desert-parsley, desert parsley, big medicine, biscuit root

WASATCH DESERTPARSLEY

L. bicolor var. leptocarpum

Other names: Gumbo lomatium

GRAY'S BISCUITROOT

L. grayi

Other names: Gray's desert parsley, Milfoil lomatium, mountain desert parsley, narrow-leaf lomatium, pungent desert parsley

Description: Biscuitroots are small perennials (most up to 10" tall, with **fernleaf biscuitroot** up to 24" tall). Typically parsley-like leaves (except **Wasatch desert-parsley** with linear leaf segments and **Gray's biscuitroot** with finely divided, thread-like segments), highly fragrant, and hollow stems. Flowers yellow to white or pink umbels, fruits flattened, winged. The plant typically yellows and withers after fruiting in midsummer. Blooms May to July. Found in dry, open foothills and montane slopes. The definitive identification for biscuitroots is its habitat of dry, low organic matter areas.

Medicinal uses: Fernleaf biscuitroot is the only biscuitroot used today by herbalists. It has been found to have antiviral and antimicrobial properties and has become an alternative treatment for AIDs and chronic fatigue syndrome. Native peoples used this plant extensively for stimulating appetite and as a tonic for weak constitutions. The whole plant, especially the root, was used for colds, flu, coughs, bronchitis, TB, pneumonia, hay fever and many chest complaints. It was used internally as a tea and in steam baths and externally for cuts, boils, bruises, acne, sores, rheumatic joints and eye infections. Native peoples used an infusion of the roots and leaves of **Wasatch desert-parsley** for chest complaints and flowers and upper leaves for colds and sore throats.

Edible uses: The biscuitroots make wonderful trail munchies. They have rather strong tasting leaves (parsley on steroids) that really refresh the mouth, and are great to slap on your turkey sandwich at lunch. Add them to salads, mixing them with other greens to tone them down. The seeds, especially green, are strong additions to stir fries. Even though their common name is biscuitroot, I find the roots hard to dig and use, and many of them are very bitter (especially the **Wasatch desert-parsley**). Some say to not dig the roots until the seed is completely ripe or they will taste like turpentine. Historically the biscuitroots were valuable food for the Native Americans, especially *Lomatium cous* which doesn't quite reach down into our area.

Garden notes: I have inadvertently planted several biscuitroots into my garden from other transplants. I love their cheery blooms early in summer, and they have found their spots in the garden by seeding around. There is nothing like munching on spring greens while working in the garden!

Caution: Some sensitive people develop skin rashes with internal consumption because of its strong cleansing action. Accurate identification of this plant is essential as it does resemble poisonous relatives like the hemlocks.

Apiaceae Family — Sweet anise — *Osmorhiza occidentalis*

Sweet anise fruit

Sweet anise habit (above) and leaves (below)

Sweet anise root

Parsley Family

Sweet Anise

Osmorhiza occidentalis

Other names: Mountain sweet cicely, sweet root, western sweetroot, western sweet cicely

Related species: Sweet cicely (*O. depauperata*) is a shorter plant, has a smaller, whiter flower, black seedheads are tapered, spreading-ascending and tend to stick to clothing. The root has a history of edible and medicinal uses. Found in forested areas of montane.

Description: Up to 3 foot tall perennial herb. Leaves twice pinnate in threes (they look like ferns!), somewhat limey-green color. Stout hollow stems. Flowers are yellowish-green in inconspicuous terminal umbels, blooming May to June. Seeds stand erect when ripe, smooth and black when mature. Roots are thick and dense; very strongly aromatic. Prefer moist, rich aspen and open forests from montane to subalpine.

Medicinal uses: The root has anti-fungal, carminative, expectorant and mild laxative properties. A root infusion or extract helps to balance blood sugar levels and inhibit fungal infections of the digestive and reproductive systems. The root is also helpful for the mucous membranes of the intestinal tract acting as a stimulant and digestive aid. A strong root infusion has been used to induce labor, treat fevers, indigestion, flatulence and stomach aches. A crushed root poultice can be used for wounds and boils. Tasty syrup can be made for coughs. Other species are generally believed to be too weak to have much medicinal action.

Edible uses: Yummy, what a treat this plant is! All parts are sweet and have an anise aroma and flavor. The strongest flavoring comes from the strange root. After a thorough washing grate the root for a soothing tea, to add to desserts (makes awesome ice cream!), or flavoring in cooking. The tea is very soothing to the stomach, much like the liqueur anisette, taken after meals to aid digestion. Dry the root and use it throughout the winter. Young leaves, flowers, and young, green seeds provide wonderful flavoring to salads and teas.

Garden notes: These are super easy to grow in rich soils in your garden by digging a small section of root. Remember to ask before you dig!

Caution: The Parsley Family has some of the deadliest herbs. Sweet anise cannot be confused with these species if you always take care to smell the foliage for the characteristic anise scent. If it does not have this, **DO NOT** use any part of the plant. Sometimes the foliage can be confused with baneberry. Again, if there is no distinctive anise scent, do not use it.

Sweet Anise Ice Cream

Use your favorite ice cream recipe with a vanilla base. Shave very finely the sweet anise root into the mixture and sample. I have used 1 tablespoon for a small batch of ice cream, but I like it strong. You will have to vary amounts to your own tastes, but start with very little then add more if you want. A little goes a long way!

Porter's lovage

Apiaceae Family · *Ligusticum porteri*

Porter's lovage habit (above) and flowers (left)

Porter's lovage seed

Porter's lovage in early spring

Porter's lovage leaf. Note that the leaf veins terminate at the outer tip of the leaf tooth. This identifying characteristic, with the strong, spicy celery-like scent of the plant and the brown hairy fringe around the top of the dark root are all critically important factors in properly identifying Porter's lovage.

Parsley Family

Porter's Lovage
Ligusticum porteri

Other Names: Osha, Porter's licorice root, lovage, fern-leaf lovage, wild parsnip, wild celery, bear root

Description: Tall, slender broadly branching perennial up to 40" tall, pleasantly aromatic leaves are large (6"-12" long) and fern-like; flowers are white in relatively small flat-topped umbels growing atop the plant; small fruits are winged and pellet shaped. Found in abundance in rich meadows and aspen forests of upper montane to subalpine. Blooms July to August.

Medicinal uses: This is still today a favorite plant of Native and Hispanic Americans. Considered a sacred plant, "bear medicine" is used in ceremonies and as medicine. With antiviral and antibacterial properties, the root, leaves and seeds are used to treat colds, flu, bronchial infections, sore throats, fever, digestive and menstrual complaints and as an infection fighting antiseptic and antimicrobial wash for wounds. Just chewing the root can help prevent infection and improve lung function. I often recommend chewing Porter's lovage, or drinking the tea, to folks adjusting to high altitudes because it helps increase respiratory capacity and efficiency. The pain associated with arthritis and rheumatism can be relieved by this plant. This is a special plant and should be used with reverence and respect.

Edible uses: The whole plant tastes like it smells— strong, spicy celery. The roots are very strong, so use the leaf and flower. It can be a bit overwhelming by itself, so mix it with other things to tone down the brashness. I love to take a leaf and put it on my ham or turkey sandwich. Mix small amounts into salads for a little spice.

Notes: The massive dark hairy root resembles bear fur, hence one of its common names "bear root." This plant seems difficult to cultivate and is becoming an example of plants we love to death; it is "At Risk," being over-harvested in many areas. Harvest with awareness and respect. Use leaves instead of roots (good medicine just a little weaker) and replant the seeds and crowns to encourage reproduction. Take only what you will really use.

Caution: This is a member of the *Apiaceae* (Parsley) Family that has poisonous look-alikes. Be sure of your identification!

Animal Teachers

Indigenous peoples have a history of deep traditional and spiritual connection with animals. They would observe qualities and traits of animals to learn from them and about the environment they share with them. They have many stories about learning the edible and medicinal uses of plants from these observations. Bears are known to have taught humans to use Porter's lovage for self-medication for arthritis and for its stimulating and anti-microbial actions. Deer and elk have been observed eating the bark of aspen during calving season. Aspen has pain relieving and anti-inflammatory properties. Zoopharmacognosy is the study of the process by which animals self-medicate by selecting and using plants, soils and insects to treat and prevent disease.

Yampah

Apiaceae Family — *Perideridia gairdneri*

Yampah flowers

Yampah fruit

Yampah habit

Yampah roots cleaned

Yampah flowers

Parsley Family

Yampah

Perideridia gairdneri

Other names: Wild caraway, Gairdner's yampah, Indian carrot, squawroot, yampa, false caraway

Description: Single, wiry stem up to 3 feet tall with linear leaves divided into 3 to 7 thread-like segments (leaves tend to wilt by bloom time), topped by a white umbel. Fruits small brown, rounded, ribbed, seed-like schizocarp. Root is a long (up to 2 inches), slender tuber-like root, one to several per plant, with dark brown covering over a white inner flesh. The aroma is like fresh carrots or caraway. Found in masses in rich meadows, aspen, open slopes in foothills and montane. Locally abundant. Blooms July to August.

Medicinal uses: The root is a mild laxative, stimulating diuretic, calming digestive aid, soothing to sore throats, helps clear mucus from the lungs and bronchials and counteracts the cathartic and emetic actions of other plants. The root poultice or wash can help draw out inflammation in wounds.

Edible uses: Where to begin with my favorite plant of the Yampa Valley? The historical uses are numerous, with the whole plant being utilized because of its sweet taste and aroma. The newly emerging leaves are wonderful in salads (mark plants in the fall to make sure you know what those new leaves look like, they are tough to see!). The flowers are also a marvelous salad addition as well as a trail munchie. The root is the most highly valued edible portion; they are sweet, tender and satisfying. They can be challenging to dig. Choose a moist day in the late summer when the leaves are yellowing (some Native American accounts say it is best after they have gone to seed), follow the stem down to the root which could be 6 to 8 inches below the surface. I do not remove the brown coating, but some people might not like the way it looks and it can be easily scraped off. The fleshy white interior is high in starch. I eat them raw in the field or put them in stir fries, soups, stews. Use your imagination on how to use them. Harrington had a great recipe for Candied Yampa Root: "Heat some butter with brown sugar and drop in whole, boiled, peeled roots. Then stir over fire until they are coated or candied. The taste may be almost too sweet for some palates." The dried roots are great throughout the winter. Hydrate them in your mouth or in water before eating or using them. The sugars and starches in yampah are rapidly assimilated, providing for an energy and endurance boost.

Notes: This is a locally abundant plant, meaning in some areas, like our beautiful Yampa Valley, this plant grows in abundance. In other areas it is scarce. Please be respectful of the plants in your area and if it is rare, use only the above-ground parts, or harvest it from your garden. Believe it or not, the state of Colorado was almost named the state of Yampa!

Garden notes: I encourage everyone to gather the dry yampah seed in the fall and strew them all over your garden (or someone else's garden!) to reenergize our valleys with this beloved plant.

"Yampa" History

Visit the mountains of Northwest Colorado and "Yampa" is a common theme: Yampa River, the town of Yampa, Yampa Valley. Yampah (yampa-riki, also known as yampa-tika) is a Ute word meaning "water plant," "common plant" or "common root." A band of Utes occupying the Yampa Valley were known as the Yamparika or Yampatika Utes for their use of the yampah plant. How fortunate we have kept this historical connection intact in our modern landscape.

Milkweed Family

Showy Milkweed

Asclepias speciosa

Other names: Milkweed, silkweed, pink milkweed, common milkweed

Description: Stout plants in patches from thick creeping rootstalks up to 3 feet tall with milky sap; leaves opposite, rather thick and grayish-green with a pinkish midrib, oblong to egg-shaped; flowers pale pink to pink or greenish-purple in umbels with 5 petals bent sharply back; fruit white hairy, often soft spiny pods with seeds tipped with fine silky hairs. Found in open moist areas along roads, ditches and valley bottoms. Blooms May to July.

Medicinal uses: Milkweed is another one of many plants with white sap used to treat warts. The sap needs to be applied daily for a few weeks for effectiveness. This plant's antiseptic sap can also be used to treat a variety of skin problems like sores, poison-ivy rash, ringworm and calluses. A fresh root poultice has been used to treat rheumatic joints. The powdered roots were used in a tea for stomach aches, coughs and venereal diseases. Use small doses only, large internal doses can be poisonous. A decoction of the tops has been used to treat blindness and snow-blindness.

Edible uses: The very young shoots can be compared to asparagus. Gather them when they are about 4-8 inches tall (much larger and they are tough and bitter), prepare like asparagus changing the water at least once to reduce the bitterness. If you miss the tender shoots, collect the youngest leaves and cook like spinach, again with a change of water and a longer cooking time than spinach. The next stage would be the flower buds gathered before they open, boiled with changes of water. The flowers and buds can be boiled down to a sugar-rich syrup. The green seed pods are boiled for an okra-like treat. Many Native Americans boiled milkweed with meat as a tenderizer. Shoshones and Cheyenne broke the milkweed stems, collected the sap and rolled it in the hand until it became firm enough to chew.

Notes: The silky hairs from the seed heads are water repellent and can be used as stuffing or woven into cloth. This floss has been used to mop up oil spills at sea. The stems provide a tough fiber for making twine, cloth and paper. The latex sap is a potential rubber source. A green dye can be made from the flowers and leaves. This is a host plant for the migrating Monarch butterfly and goldfinches use the silky hairs for nesting material.

Garden notes: There are many varieties of milkweed available for gardening. All are excellent pollinator plants. When mine bloom in the garden I just have to sit and behold the butterflies and native bees enjoying themselves.

Caution: Many members of this genus contain toxic resins, alkaloids and cardiac glycosides that have the potential to be poisonous, especially the roots. Heat destroys these toxins, so milkweeds should be thoroughly cooked before being eaten. Narrow-leaved milkweed species are the most toxic to humans. Grazing animals avoid this plant.

Plant a Monarch Butterfly Waystation with Milkweed

"By creating and maintaining a Monarch Waystation you are contributing to Monarch conservation, an effort that will help assure the preservation of the species and the continuation of the spectacular monarch migration phenomenon."

Check out www.MonarchWatch.org

Asteraceae Family   Yarrow   *Achillea millefolium*

Yarrow habit

Yarrow flower

Yarrow leaf habit (right)

Sunflower Family

Yarrow

Achillea millefolium

Other names: Western yarrow, plumajillo, staunchweed, nosebleed plant, milfoil

Description: A highly aromatic perennial with stout stems up to 30" tall from a spreading rootstalk. Leaves alternate, gray tinged, fernlike with very fine segments. White flowers in a dense flat-topped corymb-like cluster. Individual flowers small with white rays and yellowish disc flowers. Found in almost every habitat from dry to moist up to alpine; especially likes disturbed sites. Blooms June to September.

Medicinal uses: Yarrow's astringent, antiseptic, antispasmodic and hemostatic properties make it an excellent remedy for healing burns, cuts, bruises, hemorrhages, nose bleeds, inflammations, hemorrhoids and reducing heavy menstrual bleeding. The aerial parts are used. Dried and powdered, it is an excellent addition to a first aid kit to stop bleeding. As a tea, it can be used as a bitter tonic for weak digestion, to reduce hay fever symptoms, lower blood pressure, decrease menstrual pain, improve venous circulation and tone varicose veins. A tea is also helpful during colds and flu to reduce fevers by increasing sweating. A useful tea for colds is equal parts yarrow, mint and elderflower. This prolific plant is an important ally to get to know.

Edible uses: None known, except possibly as a very strong seasoning tasting something like sage. I have read where it has been used as a substitute for hops to flavor beer.

Notes: Achilles, a Greek hero, was said to use yarrow to treat his soldiers' wounds in battle, hence the genus name. During the Civil War it was used widely to reduce bleeding. The alkaloid achilleine has reduced clotting time of blood in laboratory experiments. Fresh leaves can be rubbed on the skin as an insect repellent. Yarrow tea is said to make a nice hair rinse. Flower, stems and leaves produce a yellow, gold or green dye.

Garden notes: This is another plant that will show up uninvited. Yarrow seems to grow in any habitat at any elevation. I am amazed at how prolific it is. Our native yarrow is excellent at controlling erosion in harsh areas because of its dense root growth. It is an indicator of low potassium soils. The garden yarrows have been bred for new colors and forms, and many times the valuable oils of the native were bred out. If you want to use yarrow medicinally, make sure you introduce highly aromatic plants that maintain these valuable constituents.

Caution: If you have sensitive skin, you may react to yarrow. Do not use large doses over time. Pregnant women should not use yarrow as it could cause miscarriage.

Generalists

Some plant species have very broad ecological ranges. These generalists can be found from dry to moist, shady to sunny habitats. Yarrow is probably the best example of a generalist in our area – from the four to six inch tall yarrow of the harsh alpine, to the big bushy yarrows in its prime habitat of rich soil and plentiful moisture. Since yarrow is such a valued medicinal plant, it is almost like Mother Nature is saying, "Here is my best, wherever you might need it." Many thanks indeed.

Asteraceae Family   Pearly everlasting   *Anaphalis margaritacea*

Pearly everlasting flower and leaf

Pearly everlasting habit

Sunflower Family

Pearly Everlasting

Anaphalis margaritacea

Other names: Strawflower, western pearlyeverlasting

Description: Perennial up to 2 feet tall, leafy stems with gray to white-wooly slender lance-shaped leaves with down-rolled edges and alternate on stem; basal leaves wither by flowering time. Can form extensive colonies from rhizomes. Flower heads are white with papery white bracts around yellow to brownish disc flowers, forming flat-topped clusters. Found in open, dry sites from foothill to subalpine. Blooms July to September.

Medicinal uses: Native peoples used this plant to treat a wide variety of ailments. Internally, it is a good remedy for diarrhea, dysentery and lung problems. A poultice or wash of the flowers or whole plant can be applied to burns, sores, ulcers, rheumatic joints, swellings and bruises. An infusion of the plant is steamed or smoke inhaled to treat headaches. The tea is soothing, astringent, anti-inflammatory and a diaphoretic with expectorant properties. It can be good for colds, fevers, sore throats and to expel worms. There are mild antihistamine effects that can be useful in treating asthma. The dried leaves can be smoked for inflamed mucous membranes and bronchial cough. Not used in commerce today but would be a good local medicine to add to your medicine chest.

Edible uses: The leaves and young plants can be used as a potherb.

Notes: If picked when the buds are just opening and hung to dry, these are beautiful and fragrant in dried arrangements. The leaves used alone or mixed with other plants were used for smoking. The smoke from pearly everlasting purified gifts being left for the spirits and protected houses from witches. Men chewed the plants and rubbed the paste on their bodies to gain strength, energy and protection from danger. The flowers, leaves and stems combined were used to make yellow to gold, also green and brown dyes.

Garden notes: Pearly everlasting is just starting to come into the landscape scene and are we lucky! This is a tough drought tolerant, high altitude plant with beautiful silver foliage. It can be aggressive with its spreading rootstalk, so give it plenty of room. These are valuable pollinator plants.

Why Are Leaves Different?

The myriad shapes, sizes, textures, colors and scents of leaves are astounding. Throughout the world, leaves deal with everything from torrential rain to grazing elephants, rain once every 5-7 years to ice-covered lands. Our area does not have such dramatic demands on plants but we still have a highly diversified ecology and the variety of leaves mirrors that diversity. The dry, windy, intense sun and short growing season of alpine areas have produced plants in miniature, seldom reaching over 8 inches tall, with small leaves often covered in hairs, waxes and pigments to reduce sun scald, desiccation and freezing temperatures. The cool, moist coniferous and aspen forests have larger understory plants with generally larger leaves that lack the leaf coatings of more xeric plants. Other arid areas must prevent desiccation. Leaves of these plants, like pearly everlasting and pussytoes, are small, thickened, and hairy or waxy and often turn their leaves to avoid intense mid-day sun and inroll their leaves when heavily water stressed. Gambel oak and chokecherry leaves are covered with a fatty substance, called cutin, giving the leaves a shiny appearance which helps reflect some sun and is effective in preventing water loss.

Asteraceae Family — Pussytoes — *Antennaria* spp.

Pussytoes flower

Rosy pussytoes flower & leaf

Pussytoes habit

Sunflower Family

Pussytoes

Antennaria spp.

Other Names: Catspaws, catsfoot, life everlasting, mountain everlasting, rosy pussytoes, Rocky Mountain pussytoes

Description: The identification of species can be very difficult because many species cross, mixing traits in plants, and many produce seed without pollination (apomixis). The sexes are on separate plants, and sometimes a population is equally divided with female and male, others will have only male or female. A generalized appearance would be grayish-green leaves forming low mats, flowers with disc flowers only, surrounding the heads are series of overlapping paper-like bracts of varying colors (white, brown, rosy). At the time of blooming the stems are short, then elongate up to 10 inches so the wind disperses the fluffy seeds. Found in a variety of habitats. Blooming typically mid to later summer. May be confused with the taller, broadly mounding, leafy-stemmed, strawflower-like white pearly everlasting (*Anaphalis margaritacea*), or the shorter leafy-stemmed, tend to be sticky, non mat-forming, dirty brown flowered cudweed (*Gnaphalium* spp.).

Medicinal uses: The young flowering tops in tea are astringent and diuretic. It is used for liver inflammation, irritations of the upper intestines, is valuable as a vaginal douche and is soothing for sore throats. It is said to stimulate the flow of gastric juices and pancreatic secretions and to raise blood pressure. With rich mucilage content it can be valuable for the treatment of chest complaints. The pleasant taste of the tea makes it a good choice for children's fevers and colds. A cute little plant that is often overlooked, yet an excellent medicine is hidden within.

Edible uses: The gum of the stalks can be chewed which is said to be somewhat nutritious.

Notes: The everlasting flowers are beautiful in dried arrangements.

Garden notes: Our native Rocky Mountain pussytoes (*A. parvifolia*) is a great xeric plant for open sites. They form large mats, and the silver leaves are beautiful in masses. These are tough plants, but will take extra water to get established well. Rosy pussytoes (*A. rosea*) doesn't seem to form as extensive of a mat, but the showier pink flower heads are so cute! If you are transplanting, dig smaller plants for better success. The seed takes well, but is slow growing.

Ouch!

Look at those cats, all in rows

Lying on their backs, showing their toes.

With the sun burning bright, that's bad because

Cats on their backs will get sunburned paws.

— Jack Sanders
The Secrets of Wildflowers

Asteraceae Family — Sages — *Artemisia* spp.

Prairie sage habit (above left), flower (center) and leaf (right)

Fringed sage leaf

Alpine sage habit

Fringed sage habit

Alpine sage flower

Caution: Most sages can cause allergic reactions and many are considered toxic. Always use *Artemisia* with caution. Contact dermatitis is possible in sensitive persons. Long term use should be avoided and they should not be used during pregnancy.

Sunflower Family

SAGES
Artemisia spp.

FRINGED SAGE
Artemisia frigida

Other names: Fringed mountain sage, silver sage, pasture sagewort, prairie sagewort

PRAIRIE SAGE
Artemisia ludoviciana

Other names: Western mugwort, white sagebrush

ALPINE SAGE
Artemisia scopulorum

Description: Fringed sage is a slightly woody delicate plant up to 20" tall with soft, finely dissected, silvery-hairy leaves densely massed along the stem. Abundant in dry meadows and hillsides up to subalpine.

Prairie sage forms sprawling patches of stems, differing in heights, usually more than 12" tall. Leaves highly variable in form but usually silvery gray. Most abundant in meadows and forest openings from upper foothills to subalpine.

Alpine sage is a short (up to 6" tall) perennial with finely twice-pinnate (pinnatifid with the primary divisions divided again) appearing gray-green. Flowers are dark or bi-colored dark and light. Found in alpine and upper subalpine.

Medicinal uses: *Artemisia* species all have similar properties, some much stronger than others with strong cautions on internal use. Like other members of this genus they are antibacterial, antifungal, antiparasitic and very bitter. (This species is not related to garden sage, *Salvia*.) Bitterness stimulates digestion by increasing gastric juices and is used as a digestive aid. Local species were used extensively by Native peoples to treat a variety of problems. Leaf infusions were used for colds, flu, coughs, sore throats, fevers, lung problems, diarrhea, Rocky Mountain spotted fever, to expel worms and in bathwater for arthritis. A leaf poultice was used for sores, bleeding wounds, blisters, burst boils, sore eyes, insect bites and itching. Leaves crushed into a snuff were used for nosebleeds, headaches and sinus problems. Today, extracts of *artemisinin* from other *Artemisia* species are used in the treatment of malaria. **Prairie sage** was called "women's sage" because of its use to treat menstrual problems. Smoke from burned leaves was inhaled for chest complaints. I can find no record of use for **Alpine sage,** but that is probably due to the much wider availability of the other species.

Edible uses: All *Artemisia* listed here are highly aromatic and can be used sparingly as a seasoning. The seed-like fruits can be eaten fresh, dried or pounded into a meal.

Notes: Fringed and Prairie sage were used as insect repellants, room disinfectants, underarm and foot deodorants and for purification. They have a long history of ceremonial use, especially in sweat lodges and for the Sun Dance. They were both used as a veterinary medicine for horses. They produce a green dye. Shredded bark of **Prairie sage** was used for candlewicks and as fiber to produce footwear, clothes and sacks.

Garden notes: Fringed sage is one of the loveliest filler plants for any xcric garden. The soft foliage placed either singly or en masse really helps accentuate colors in the garden. They can be somewhat spindly so I like to put at least three plants together for a fuller look. Do not grow in a rich soil with extra moisture or your plants will be lanky and short-lived.

Asteraceae Family Balsamroot *Balsamorhiza sagittata*

Arrowleaf balsamroot habit.
Flower and leaf (inset)

Sunflower Family

Arrowleaf Balsamroot

Balsamorhiza sagittata

Other names: Arrow-leaved balsamroot, balsamroot

Description: Spectacular sunflower-like flowers rise singly above a clump of velvety silver leaves. The large arrowhead-shaped leaves are almost all basal. Each plant comes from a large, very aromatic and resinous taproot. Typically found on dry, hot hillsides of the montane. Blooms May and June. These can be confused with mules's ears (*Wyethia amplexicaulis*), but mule's ears have lanceolate leaves that are green.

Medicinal uses: The roots are anti-microbial, diaphoretic, disinfectant, decongestant and mildly stimulating to the immune system, a local substitute for Echinacea. The tough outer skin of the root must be removed. Among Native Americans the leaves, roots and stems were used as a poultice or infusion for minor wounds, poison ivy, insect bites, burns and as a treatment for stomach pains, colds, whooping cough, tuberculosis (TB), fevers, labor pains and headaches. The smoke from the root was inhaled as a remedy for bodily aches like rheumatism. I combine this plant with our *Rudbeckia ampla* as a good remedy for colds and flu and whenever immune system stimulation is needed.

Edible uses: This plant contains a hard to digest sugar that makes the roots very bitter. Slow cooking converts the bitter sugar to fructose. The large starchy taproot can be peeled, boiled and ground, then cooked and dried to make a meal. The peeled roots can also be barbecued, sliced and sautéed, roasted, or eaten raw (very bitter and rough on the stomach!). The root can be considered a survival food as it is available year round. Young flower stems can be peeled and eaten like celery. The young spring leaves can be cooked like asparagus, or used raw in salads (very strong flavor). The seeds were made into a tasty cereal by heating the seeds on hot coals, winnowing the seed, then grinding. This meal can be mixed with boiling water to make a mush that tastes somewhat like popcorn. The gum from the roots can be made into a chewing gum.

Notes: The root was burned as incense. The large hairy leaves were used as insulation in shoes. This perennial plant is very long-lived, not flowering before the 4th or 5th year and continuing for over 50 years. Because of this, be conservative when harvesting, and avoid steep hillsides where this plant performs an important role in protecting against erosion.

Garden notes: Arrowleaf balsamroot is very difficult to grow because of its massive taproot. Once established it will be there for a long time! If transplanting, dig only the youngest plants found. Seed sown in the late fall is fairly successful if the critters don't get it.

Caution: Do not confuse Arrowleaf balsamroot with the poisonous leaves of mule's ears.

"Be like the flower, turn your face to the sun."

— Khalil Gibran

Asteraceae Family | Chicory | *Cichorium intybus*

Chicory flowers

Chicory basal leaves

Chicory habit

Sunflower Family

Chicory
Cichorium intybus

Other names: Wild succory, blue-sailors, coffeeweed

Description: A weedy perennial up to 4 feet tall. Early spring produces a rosette of deeply toothed basal leaves, similar to dandelions, except chicory leaves are somewhat hairy. The leaves sprout from a deep stiff, beige taproot. The flower stalk is somewhat hairy with alternate, lance-shaped, toothed leaves, with leaves reduced toward the top of the plant. Sky blue flowers are stalkless along the stem, from the leaf axils, with ray flowers only. Only a few bloom at once. Found in disturbed sites. Blooms spring through fall.

Medicinal uses: The first written record of medicinal use of chicory is in Egypt in 4000 BCE. It has also been identified as one of the bitter herbs in the Bible. 1800 years ago, the Roman Galen proclaimed it "the friend of the liver." Chicory is an excellent bitter tonic for the liver and digestive tract. It is therapeutically similar to its cousin dandelion, supporting the action of the stomach and liver and cleansing the urinary tract. The roots and leaves are used, fresh or dried, the roots being more active medicinally. Chicory is recommended for loss of appetite and dyspepsia and is also taken for rheumatic conditions, jaundice, liver enlargement and gout. The root is a safe diuretic, increasing both the water and waste products in the urine. An infusion of the leaves and flowers also aids digestion. A fresh plant decoction may alleviate gallstones and kidney stones and aid in the production of bile. Research has shown the root to decrease the heart rate and amplitude indicating its potential for use in heart irregularities. It has also been shown to significantly lower cholesterol and blood sugar levels and the roasted root has antibacterial properties. The latex in the stems is applied to warts in order to destroy them. A weed, yes, but such a useful one!

Edible uses: Collect the young leaves in early spring for salads or a potherb. They can be rather strong. Use the greens until the flower stalk appears in mid-spring. The edible flowers wilt soon after they are picked and are not very tasty, but because they are available all summer, try dipping them in batter and frying them. The taproot is very bitter and tough; try roasting it at 250°F for 2-4 hours until it is dark brown, brittle and fragrant. This can be used as a coffee substitute by grinding it in a blender and brewing it like freshly roasted coffee. It has a mild coffee flavor without the over stimulating caffeine of coffee. Very young taproots are much milder and are cooked like parsnips. The roots are also used in seasoning soups, sauces and gravy, imparting a deep rich color.

Notes: Chicory is on the Colorado Noxious Weed List C. A blue dye is obtained from the leaves. The plant is used in Bach flower remedies — keywords for prescribing it are 'Possessiveness,' 'Self-love' and 'Self-pity.' The flowers are an ingredient in a commercial herbal compost activator. The starch 'inulin' in the roots can easily be converted to alcohol, which has a potential to be used for the production of biomass for alternative fuels.

Garden notes: Chicory found in gardens indicates soils with clay texture, with high humus or fertility unless the weeds are pale and stunted, then fertility is low.

Bird Enticers
Grow this beautiful "weed" in your yard and as it goes to seed you will be sure to have an abundance of feathered visitors. American goldfinch love this plant, as well as pine siskin and many other seed-eaters.

Asteraceae Family — Gumweed — *Grindelia squarrosa*

Gumweed milky buds & flower (above); serrated leaf clasping stem and often characteristic red stem (both insets left)

Gumweed habit

Sunflower Family

Gumweed

Grindelia squarrosa

Other Names: Curly-cup gumweed, rosin weed

Description: Biennial or short-lived perennial with branched stems up to 2 feet tall. Highly aromatic leaves alternate, often clasping the stem, oblong, somewhat lobed. Yellow flowers with 25-40 rays, yellow disc flowers, and several layers of overlapping backwards curling, sticky bracts below the flower head. Found in dry and disturbed sites from foothills to montane. Blooms July to September.

Medicinal uses: The flowers and resinous buds of gumweed can be used as a tea, tincture, syrup or lozenge for respiratory conditions like bronchitis, colds, coughs and especially asthma when an expectorant is needed. A poultice or wash of this tissue healing plant can be applied to poison ivy, measles, wounds, dermatitis and various skin irritations. The buds in tea or tincture form can relieve indigestion, colic and stomachaches, has diuretic properties and has been used to treat kidney problems. Historically, this plant was used to treat a wide range of other ailments including headaches, malaria, gonorrhea, pneumonia, smallpox, tuberculosis (TB), and cancers of the spleen and stomach.

Edible uses: None of note.

Notes: Buds and leaves used to make a wash for healing saddle sores on horses. Yellow and green dyes can be obtained from the flowering heads and buds.

Garden notes: Although most people consider gumweed a "weed," it is a valuable plant for nasty xeric sites. I have taken seed and thrown it over gravel areas and other poor soil sites and the plants seem to love it. It can be short-lived, so seed might need to be reintroduced.

Caution: Grindelia may take up and store selenium from the soil so large doses can be mildly toxic. Its high resin content makes it hard on the kidneys, so is best used only for acute ailments. Small doses of grindelia are thought to lower the heart rate.

Disturbed Habitats and Weeds

Mother Nature prefers diversity. Where there is bare soil, she will fill it, usually with what we call "invasive weeds". Weeds have advantages that allow them to proliferate in disturbed environments. They grow quickly and often produce large numbers of seeds that can survive in the ground for many years. Many have underground or aboveground stems that spread out and "take over." They flourish in soils that are depleted, polluted and compacted. These same advantages that favor their growth are creating biomass, preventing erosion, bringing up nutrients, revitalizing the soil, attracting beneficial insects and feeding the soil food web. If it weren't for weeds, the world would have lost more topsoil than it has to date. Research is showing us that many weeds have the capacity to clean poisons and heavy metals from contaminated soils. Weeds are excellent adapters and perform a vital job in disturbed ecosystems. Although at times their aggressiveness can reduce native diversity, that loss of diversity could be due to our own destruction of natural habitats. Accidently and purposefully, we have a large part to play for their presence.

Asteraceae Family — Little sunflowers — *Helianthella* spp.

Five-nerved sunflower habit (left) and flower (above)

Little sunflower flower (below) and habit (right)

Sunflower Family

LITTLE SUNFLOWERS
Helianthella spp.

FIVE-NERVED SUNFLOWER
Helianthella quinquenervis

Other names: Aspen sunflower, nodding sunflower, fivenerve helianthella

LITTLE SUNFLOWER
Helianthella uniflora

Other names: One-flowered little sunflower

Description: Both Little sunflowers are tall, lanky plants. **Five-nerved sunflower** grows up to 3 feet tall and **Little sunflower** grows a little shorter and more "wimpy" looking. Both have the typical yellow sunflower head, but these nod always to the east, and are usually solitary on the stem (**Five-nerved sunflower** can have some smaller heads below). Ray flowers are yellow; disk flowers are deep yellow to brownish-red. Both have large stem leaves. **Five-nerved sunflower** only has enlarged basal leaves with highly defined set of five veins along the leaf. **Five-nerved sunflower** is very common, typically higher elevation, foothills to subalpine, in aspen, open rich meadows and hillsides. **Little sunflower** is at lower elevations of the montane in sagebrush and open meadows. Both bloom July through fall.

Medicinal uses: There is record of Paiute and Shoshoni peoples using the hot, mashed roots as a poultice for swellings, sprains and rheumatic pains. An infusion of the roots was used as a wash or compresses for headaches. I can find no information of any internal uses. I have no personal experience using these plants.

Edible uses: The highly touted *Helianthus* spp. definitely have larger seeds, but both of our little sunflowers are just as tasty. Try roasting them for more flavor.

Garden notes: These are stunningly beautiful, long flowering treats for your garden. I have transplanted the Little sunflower into my dry berm, nesting it into meadow grasses. Five-nerved sunflower was transplanted into the shadier section of my garden. Both are thriving, providing rich nectar for pollinators and seeds for the birds. What a treat!

Dyes from Plants

Plants have long been a source for natural dyes. Flowers in bloom, ripe berries and nuts, roots and barks can be used. Remember to follow ethical harvesting guidelines when gathering plant material for dying. Here are some examples of colors and the local plants that can provide them:

Yellow, brown, and black: alder, oak, usnea, dandelion, goldenrod, sagebrush, mullein, sunflowers, rudbeckias, willows, rabbitbrush, lupine, stinging nettle, red clover, Douglas fir, yellow dock, pearly everlasting, aspen, kinnikinnick, pines, juniper, Oregon grape, hops, and wild rose.

Red, purple, and blue: red osier dogwood, bedstraw, dandelion, chicory, elkbrush, larkspurs, horsetail, western stoneseed, chokecherry, thimbleberry and huckleberry.

Green: stinging nettle, lamb's quarters, yarrow, milkweed, rabbitbrush, pine, self-heal, and wild rose.

Asteraceae Family — Wild lettuce — *Lactuca serriola*

Wild lettuce flower

Wild lettuce habit

Prickly midrib on underside of clasping wild lettuce leaf

Sunflower Family

Wild Lettuce

Lactuca serriola

Other names: Prickly lettuce, compass plant, opium lettuce, lettuce opium, China lettuce

Description: Alien biennial from a taproot with milky juice, up to 3 feet tall, branching only in flowering portion of the stem. Leaves are alternate, pinnately lobed or lobeless, prickly on the lower side of the midrib, clasping the stem. Flower heads yellow with ray flowers only. Common in disturbed sites. Flowers from June to September.

Medicinal uses: Sometimes called lettuce opium because the sap is reminiscent of the milky white latex from the opium poppy. The sap flows freely from any wounds, hardens on contact with air and is very tedious to collect. I opt for using the above ground flowering plant as a fresh plant tincture or dried for tea and combined with other herbs. It does have a very mild analgesic effect without digestive upset or addictive properties, safe enough even for children. In children, it can calm overstimulation and excitability. It is a mild sleep aid and pain reducer. It has some antispasmodic properties useful for whooping cough, dry, irritated coughs and menstrual and intestinal pains. The juice from the stems makes an astringent skin lotion.

Edible uses: Collect the leaves as soon as the rosette appears in early spring. Use these leaves like dandelion or chicory, raw or cooked. They become extremely bitter after the flower stem emerges. The flowers can be used as dandelions are used.

Notes: Wild lettuce is cultivated in Egypt for the seed which contains a semi-drying oil that is used in soap making, paints, varnishes, etc. As a compass plant, the top leaves align north-south.

Garden notes: Wild lettuce will eventually make it into your garden via its airborne seeds. It easily crosses with domestic lettuce creating a not-so-tasty lettuce hybrid.

Caution: Large doses of the sap may cause drowsiness, restlessness and loss of sex drive. Overdoses can cause death through cardiac paralysis.

Soil Condition Indicators

What might our weeds be telling us? Can it be that they are reliable indicators of potential fertility problems? A few weeds might not mean much but consistent populations of the same species, especially perennials, can indicate soil conditions. Yarrow indicates low potassium while knapweed and red clover can indicate excess potassium. Plantain, Canada thistle and dandelions indicate heavy clay soils. Lots of healthy leguminous plants like clovers or stunted, yellowish non-leguminous weeds are a sign that soil nitrogen is low. Mullein is found in uncultivated soils with low fertility. Wild lettuce, lamb's quarters, and nettles can indicate soil that has been tilled. The docks (*Rumex* spp.), horsetail and willows, rushes, sedges and goldenrod can indicate waterlogged or poorly drained soil. Chicory and bindweed indicate hardpan, compacted soils. Lots of the mustard family weeds (*Brassicaceae*) can mean soil rich in phosphorus. An internet search offers many useful articles and charts on this topic. Use this information to plant garden plants that thrive in similar conditions to what the weeds indicate or amend the soil so conditions are less inviting to the weeds. Most of all, listen to the weeds!!!

Asteraceae Family — **Pineapple weed** — *Matricaria discoidea*

Pineapple weed leaf and flower

Sunflower Family

PINEAPPLE WEED

Matricaria discoidea
(*Lepidotheca suaveolens*)

Description: A non-native annual which is decumbent up to 1 foot tall with leaves greatly divided into fine, short segments. The heads are cone-shaped, yellowish green, lacking ray flowers, each head surrounded by several overlapping bracts with papery margins. The plant gives off a pleasant "pineapple" odor when crushed. Common on disturbed sites. Mayweed chamomile (*Anthemis cotula*) is similar in appearance but taller, with white ray flowers and a disagreeable odor.

Medicinal uses: If it has a smell (pineapple or chamomile like), it has some medicinal properties similar to chamomile. It is a mild sedative with stomach calming actions. It can also be useful to treat fevers, colds, intestinal worms, diarrhea and menstrual cramps. Native peoples used it as a childbirth aid, for infant convulsions and mixed with other aromatic plants as a perfume.

Edible uses: An infusion of one tablespoon of fresh leaves, stems and flowers in a cup makes a refreshing chamomile-like beverage. The flower heads can be added to breads and muffins, salads and eaten as trail munchies (very strong flavor!).

Notes: The dried flowers can be used as an insect repellent and they were used to line cradles and stuff pillows.

Garden notes: The plant repels insects, and the dried flowers are used as an insect repellent in the garden. The presence of Pineapple weed in our gardens indicates hardpan or crusty soils.

Caution: Folks who are allergic to other members of the Aster family may be allergic to this one.

Talking "Plant" to Someone in Russia, Kenya or Anywhere!

Scientific names are very descriptive. Most of these names are Latin-based although Greek and other languages are used. These descriptors can often have a prefix or suffix which show size or position of something, e.g., *macro* means large or long, so *macrophyllum* is a large leaf (*phylum* meaning leaf) and macrostylus means long styled (style being the female part of the plant). Other names denote the general **"personality"** of the plant: *bellus* (beautiful), *callistus* (very beautiful), *elegantissima* (very elegant), *robustus* (growing strongly). **Color** is often found in scientific names: *alba* (white), *argenteus* (silvery), *niger* (black), *roseus* (rosy red), *ruber* (red), *coeruleus* (blue), *lividus* (grayish blue), *purpureus* (purple), *aureus* (golden), *flavus* (bright yellow), *luteus* (yellow), *glaucus* (sea-green), *sempervirens* (always green), *viridis* (green). **Markings** on the plant surface are noted in many ways: *maculatus* (blotched or spotty), *pictus* (painted). **Shape** of the flower, leaf or plant are denoted: *racemosus* (flowers in a raceme), leaves can be *crispus* (curled), *dentatus* (toothed), *digitalis* (fingered, like digits), *linearis* (narrow), *pungens* (pointed), the shape of the plant can be *cespitosus* (dense clumps), *gracilis* (graceful, slender) or *reptans* (creeping). **Fragrance, taste and use** include *edulis* (edible), *fragrans* (fragrant), *graveolens* (strongly fragrant), *irritans* (irritable, discomforting), *odoratus* (fragrant), *officinale* (used in medicine), *phu* (stinking), *pungens* (pungent), *sativus* (sown or cultivated), *suaveolens* (sweet-smelling). Other identifying factors include texture, size, direction of plant growth, flowering time, habitat, something the plant resembles and often people.

"Without names, no knowledge."
— Linnaeus

Asteraceae Family Rudbeckias *Rudbeckia laciniata* var. *ampla*

Goldenglow habit, leaf and flowers

Sunflower Family

GOLDENGLOW

R. laciniata var. ampla
(*Rudbeckia ampla*)

Other names: Wild golden glow, tall coneflower, cut-leaf coneflower

Description: A stunning perennial up to 6 feet tall, branching stems with alternate leaves, pinnately divided into 3 to 5 broad sections. Ray flowers are yellow, up to 2 inches long tending to slightly reflex from a cone-shaped mound of dull green or greenish-brown disk flowers. Found by streamsides and rich, moist forest openings in montane. Blooms July to August.

Medicinal uses: This is a stimulating diuretic, with mild cardiac stimulation as a side effect. It is anti-microbial and has immunomodulating properties. The root has a history of use for painful menstruation. This species was generally considered a kidney medicine and used as a tonic, diuretic and soothing balm for long term inflammation of the mucous membranes of the urinary tract and for indigestion and colds. A poultice of the flowers was used to treat burns and a root infusion used as a wash for sores, swellings, and snakebites and as ear drops for earaches. Currently, the roots are employed as an analog and substitute for echinacea, acting to stimulate the immune system while removing the waste products produced. I have found goldenglow combined with another echinacea substitute, arrowleaf balsamroot (*Balsamorhiza sagittata*), to be an excellent local remedy for colds and flu and where immune stimulation is needed.

Edible uses: This plant is considered not highly edible because it is bitter and astringent.

Notes: Green or yellow dyes are obtained from the flowers.

Garden notes: Even though goldenglow is typically found in moist, rich sites in nature, I have it growing in a south-facing bed with richer soil and it seems to thrive there. I think a little dryness might even keep it from being so floppy. There are many *Rudbeckia* on the market now, but the true natives are a little more difficult to locate.

Caution: Not for use during pregnancy. People sensitive to other members of the Sunflower family may also be sensitive to Rudbeckias. All are said to poison cattle, sheep and pigs.

Rudbeckia's Connections

Jack Sanders, in his book *The Secrets of Wildflowers*, provides an interesting story about our beloved *Rudbeckia* genus. Two Swedish botanists in the 18th century, Olaus Rudbeck Sr. and Jr., were highly respected professors at the University of Upsala in Sweden where a rural minister's son, named Carl von Linne, studied medicine in the 1720s. Rudbeck Jr. admired von Linne's interest in botany and von Linne moved into Rudbeck Jr's house, earning money by tutoring 4 of the 24 of Rudbeck Jr's children (wow!).

Von Linne went on to develop our modern day binomial naming system for all living things using genus and species, and he later renamed himself Linnaeus. It is not clear why, but he decided to name these North American flowers after these mentors from his youth.

Asteraceae Family — Goldenrods — *Solidago* spp.

Giant goldenrod habit and flower (upper right)

Rocky Mountain goldenrod habit

Sunflower Family

GOLDENRODS
Solidago spp.

Other names: Golden woundwort, fastening herb

Description: There are many species of goldenrod in our area, from small alpines to the 5 foot *S. gigantea*. All goldenrods have many small heads with short rays. Both the ray and disc flowers are yellow. In some species the heads are concentrated on one side of the inflorescence branches. They have somewhat narrow leaves with small teeth. Most goldenrods are late summer bloomers, found in a variety of habitats from foothills to alpine.

Medicinal uses: The older names of this plant indicate its long standing reputation as a wound herb used before battles as a precaution against infection and bleeding. It was used as a powder, tea or poultice for wounds, burns, ulcers, bruises, and rheumatic and nerve pain. Today it is also recognized as a safe and gentle remedy for many problems affecting the urinary and respiratory systems. It is used for colds, flu, sore throats, skin problems, upper respiratory infections, kidney and bladder infections and stones. It also appears effective in treating the Candida fungus which causes yeast infections and oral thrush. Some herbalists use goldenrod before hay fever season to strengthen allergic defenses and build resistance to seasonal allergens. There is some research that this plant has hypoglycemic activity that may help with diabetic conditions. A homeopathic remedy of the plant is used to treat kidney and bladder problems, rheumatism and arthritis. This herb is also a gentle herb for gastrointestinal problems in children.

Edible uses: The dried leaves and fully expanded flowers produce an enjoyable tea. Flowers are used as a salad garnish, and seeds thicken stews and gravies. Young greens can be cooked and eaten like spinach, or added to soups. The palatability varies depending on the habitat, age, and your taste buds.

Notes: Its scientific name *Solidago*, Latin "*solidus*" and "*ago,*" translates "to make whole" because of its medicinal properties. The sap is high in latex and has been cultivated in the past for a domestic supply of rubber. Goldenrod galls can contain larvae which make great fish bait. Flowers are used to create a yellowish-tan (with alum mordant) or golden dye (with chrome mordant).

Garden notes: Goldenrods are gaining popularity in the garden for their late season showy displays. The native species can be quite aggressive, so unless you have plenty of room, the newer cultivars might be better for smaller gardens. Native species are easily transplanted and it's usually best to do this in the spring.

> **Caution:** Some people believe they have an allergic reaction to the pollen, but most of the allergic reactions are probably due to ragweed often growing nearby. Goldenrod pollen is very heavy and is carried by insects and not wind, like ragweed pollen.

> "People from a planet without flowers would think we must be mad with joy the whole time to have such things about us."
>
> — Iris Murdoch

Asteraceae Family — Dandelion — *Taraxacum offininale*

> "Dandelions are the supreme symbol of the failure of human control, a yellow flag of mockery, and every time we burn that flag, back it comes, stronger than ever. No plant or animal is as obstinately perverse in its taunting of human wishes."
> - David Ehrenfeld

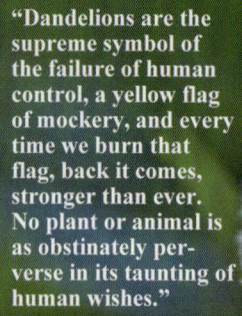

Dandelion fruit

Dandelion flower, leaf & habit

Garden notes: Dandelion is working in your garden to attract ladybugs, aerate the soil and provide early spring pollen for bees and other pollinators. The presence of dandelions in your garden indicates heavy clay soils, acidic or low lime situations. No need to plant these, every garden has its regular volunteers, so use them! To keep your neighbors happy, be sure to mow off the flowers just before they go to seed.

Caution: Be sure to harvest in 'clean' areas not sprayed with chemicals or not visited by pets. Dandelion's strong diuretic action can lower blood pressure so folks with low blood pressure should not use dandelion on a regular basis.

Sunflower Family

Dandelion

Taraxacum officinale

Other names: Lion's tooth, swine's snout, Irish daisy, puff ball

Description: Alien perennial herb from a thick taproot up to 16" tall. Basal rosette of leaves are variously pinnately lobed, long and slender. Yellow flowers in single heads per leafless, hollow stem; ray flowers only producing seed-like achenes topped with a "parachute" of white hairs. Large, stout taproot with bitter milky sap. Found everywhere, pretty much.

Medicinal uses: The entire plant is considered a whole body tonic and one of the safest and most important herbs for the liver. Dandelion roots, gathered in the early spring or fall, are mildly laxative and diuretic, an excellent liver and stomach herb used to increase the production of liver secretions such as bile, and strengthen the liver's ability to clear toxins out of the blood, resulting in a clear complexion, better digestion, less bloating, sounder sleep and improved appetite. The unroasted roots are the most medicinal and the root tea, tincture or capsules have been used to treat such problems as jaundice, hepatitis, hypoglycemia, muscular rheumatism, eczema, acne and anemia. Regular use of dandelion can strengthen and tone the entire digestive system. Dandelion leaf has been shown to relieve PMS, vaginal dryness, and arthritis, as well as improve circulation and treat urinary problems. The leaves are a good diuretic for fluid retention and eliminating toxins and treating constipation. The young spring greens picked before flowering, eaten a few times a week for about six weeks can promote spring liver cleansing. Once they have bloomed, their taste becomes too bitter for some folks. Younger leaves throughout the summer and in the fall (after the first frost) continue to be usable. The bitter taste in foods stimulates gastric juices and helps to improve digestion. The blossoms can be infused in oil and used in a muscle relaxing massage oil or salve. A blossom tea or facial steam can help heal chapped or wind burned areas, age spots, large pores, wrinkles and bring the skin chemistry back into balance. Regular application of the milky sap is said to kill warts. This is a valuable 'weed' worth getting to know!

Edible uses: Where to begin with this ubiquitous sunny plant? The early spring leaves are fantastic in salads or lightly steamed or stir fried. Older leaves are very bitter; the sharp taste can be lessened by cooking the leaves in salted water and adding a small amount of vinegar to the finished dish. I can use the plants growing on the north side of my house virtually all summer as they are not made bitter by the sun and dryness. Use the just emerging buds in salads or stir fries and the open flowers as fritters by dipping them in pancake batter and frying them. The pretty ray flowers are nice in salads, add color to butter and make a beautiful wine. The roots are strong and bitter, tasting somewhat like parsnips when boiled in at least 2 changes of water. Some people place a pinch of baking soda in the water during cooking. The roots are roasted slowly until dark brown and used as a coffee substitute (with a large amount of imagination). The seeds, minus the fluffy tops, were eaten as a trail munchie, ground into flour or used to grow sprouts.

Notes: The flowers produce a yellow dye and the roots a magenta color. Dandelions don't necessarily need other dandelions to reproduce. They can reproduce via a process called apomixis — they develop seeds without fertilization. An evolutionary advantage if you are trying to take over the world! The Latin *officinale* means historical use as a medicine.

Asteraceae Family — Salsify — *Tragopogon* spp.

Meadow salsify flower

Meadow salsify habit (above left)

Yellow salsify flower

Salsify fruit

Sunflower Family

SALSIFY
Tragopogon spp.

YELLOW SALSIFY
T. dubius

Other names: Common salsify, oysterplant, goatsbeard, wild salsify

MEADOW SALSIFY
T. pratensis

Other names: Jack-go-to-bed-at-noon

Description: Both species are alien biennials (occasionally perennials) 1-3 feet tall, more or less branched, arising from a long taproot. Alternate leaves are grass-like, clasping the stem. Single flower heads atop long, hollow peduncles consist of all ray flowers, subtended by long slender bracts. Fruit is an achene, with a fluffy parachute, these forming a huge ball of fuzz. All parts contain a milky sap.

Yellow salsify flowers are pale lemon yellow, all shorter than the phyllaries; plants are found in dry hot valleys.

Meadow salsify has chrome yellow flowers that equal the purple margined phyllaries, growing at higher altitudes in more mesic meadow sites. These two species hybridize readily. Possibly found migrating from lower elevations into our areas.

Purple goatsbeard (*T. porrifolius*), a purple flowered variety.

Medicinal uses: Our common yellow species have some history of use as a ceremonial treatment for throat problems, in lotions for boils and infusions to treat internal injuries in horses. The root was sometimes used as a remedy for upset stomach and other gastric disorders. It is little used today by herbalists. **Purple goatsbeard** was used as a ceremonial emetic by Native Americans. This plant was used in a lotion for mad coyote bites on humans and livestock. The white latex was chewed like gum. It has cleansing actions helpful for the liver and gallbladder. The root was a folk remedy for treating obstructions of the gall bladder and jaundice. I do not know if the yellow species was or can be used similarly.

Edible uses: Collect the very young leaves (usually in rosettes in early spring) and the crowns in spring. They become bitter as soon as the stem shoots up. Cook as a potherb. Collect the large taproots in fall or in early spring when still a rosette (mark these plants in the fall as the grass-like leaves are difficult to identify). The roots taste like bitter oysters (use your imagination). Use them immediately or store them in sand in a cool area.

Notes: Pappus of the fruits was used to make "cotton" for stuffing pillows.

Garden notes: Purple goatsbeard has been long cultivated in gardens for the tasty roots.

Historic Recipes

Salsify Salad (from 1885)
"Wash, peel, and boil eight salsify roots. Drain them, and when they are cold, slice them into small stems about two inches long. Put them in a salad bowl, season with salt, pepper, sweet oil and vinegar, and add a few finely-sliced gherkins. Mix them all well together, dress them nicely, and sprinkle a little finely chopped parsley over the salad."

Mrs. Caroline N. S. Herbemont's Fried Salsify (1830)
"1st, Boil the Salsify, scrape them, cut them in half, length-wise, and dip them in a rich batter, and fry them in lard. In making the batter, put in a large teaspoonful of ground ginger."

— **American Heritage Vegetables** http://lichen.csd.sc.edu/vegetable/index.php

Mule's ears

Asteraceae Family — *Wyethia amplexicaulis*

Mule's ears habit, (above and right)

Mule's ears leaf

Sunflower Family

Mule's Ears

Wyethia amplexicaulis

Description: Perennial plants up to 2 feet tall, with the majority of the plant being the huge, mostly basal, glossy looking, strap-shaped leaves (like a mule's ear!). Stem leaves are reduced. The resinous leaves are highly aromatic. Bright yellow sunflower-like flowers up to 6 inches across, disk flowers yellow. Very large taproot. Forms extensive colonies in dry fields, meadows and open forests from montane to lower subalpine. Blooms June.

Medicinal uses: Native peoples used the mashed root externally as a poultice for arthritis and rheumatic pain, swellings and bruises. Navajo and Hopi found that the root of this plant acted as an emetic, though this use may be dangerous. I have no personal experience with this plant as medicine but I usually don't use such strong plants and prefer to work with those a little gentler.

Edible uses: The large taproots were fermented on heated stones in the ground for a couple of days. The seeds, though very small, can be used as other sunflower seeds. They are a little blander.

Notes: The yellow mule's ears often hybridize with the white mule's ears, producing a pale yellow-flowered plant with peach fuzz on the leaves. I have also found hybrids of a cross of probably *W. arizonica* and *W. amplexicaulis* at lower elevations in the valley. These leaves are very hairy, but look, otherwise, like our *W. amplexicaulis*. The presence of large stands of mule's ears on ranchlands often indicates overgrazing.

Garden notes: I personally have tried repeatedly, and never had luck, at transplanting and seeding these tough, gorgeous plants. I had a little success transplanting tiny seedlings just starting to grow. Let me know if you have found the magic bullet!

Caution: The leaves should never be taken internally as they are considered poisonous.

Age of Perennials

Age of plants is calculated using various techniques from tree ring counting to tagging plants and recording how many growth stages it proceeds through and how many years it spends in each growth stage. This information is rather scarce because long-term studies are rarely conducted and plants often outlive the research project. Some perennial plants are extremely long-lived, often persisting for 20 or more years. Some of these grow from a rhizome system that produces clones. We have many examples of native long-lived perennials: arrowleaf balsamroot, five-nerved sunflower and mule's ears are estimated to live 40, 40 and 28 years respectively. This is, of course, assuming habitats have not been disturbed by grazing, development, herbicide use, etc. When harvesting sustainably, the age of individuals should be considered. Harvesting without regard to age and reproductive capacity can result in weakening the gene pool. Take young plants rather than older, sexually mature plants. This is why we say to "leave the grandmother and grandfather plants" when harvesting. In our home landscapes, planting natives and long-lived perennials is also a way of preserving important species for future generations.

Boraginaceae Family — Hound's tongue — *Cynoglossum officinale*

Hound's tongue fruit

(left) Hound's tongue fruiting habit

Hound's tongue flower

Hound's tongue habit

Borage Family

Hound's Tongue

Cynoglossum officinale

Description: Alien biennial growing up to 3 feet tall from a taproot, with a rosette of simple, lanceolate, rough-textured leaves from 1 to 12 inches long. Rosettes form in the first year with a central stem producing flowers in the second year. Flowers reddish-purple, on long 1-sided spreading branches. Fruits prickly nutlets that attach to clothing and fur. Found on disturbed ground. Blooms May to July.

Medicinal uses: The name officinale indicates there is a long history of medicinal use of this plant, although because of concerns about the toxicity of certain constituents, it is rarely used internally by herbalists today. I use the leaves in salves to treat wounds, minor injuries, bites and ulcers— the same way I would use its cousin, comfrey.

Edible uses: The young leaves have been boiled and eaten, but ingestion of any part is not recommended.

Notes: Hound's tongue is on the Colorado Noxious Weed List B. The beautiful purple flowers produce many, many sticky nutlets that "hitch-hike" on pet fur and socks and are easily spread throughout the landscape. Since it is a biennial, it is possible to control this plant in your gardens and fields by preventing seed production.

Caution: Hound's tongue contains alkaloids that can cause cancer when the plant is consumed in large quantities. The plant is also said to be slightly poisonous; there are no reported cases of human poisoning but there are many cases of livestock being poisoned. The plant has a disagreeable odor and taste so is seldom eaten by animals. Contact with the plant can cause dermatitis in sensitive people.

Seed Dispersal

Plants want their seeds to sprout and grow far away because their young plants are future competition and, since plants can't move, the older plants are stuck with the onslaught. So seed dispersal is beneficial to both parent and offspring in reducing competition, possibly expanding the range of the plant into new environments and, in the case of many animal dispersed seeds, can come with an added benefit of fertilizer. There are 5 seed dispersal mechanisms: gravity (think coconuts falling), wind (dandelion parachutes and maple winged seeds), water, ballistic (our native geraniums are fascinating in catapulting their seeds when the capsules have dried) and by animals. And here we are at hound's tongue, the ultimate animal dispersed seed. In fact, many of our "weedy" species are animal dispersed seeds similar to hound's tongue as they can be carried longer distances. Animals disperse seed through "hitchhiking" like hound's tongue, or in a gut. After the animals enjoy the chokecherries, serviceberries or rose hips, it is "planted" with some fantastic fertilizer.

Brassicaceae Family | Shepherd's purse | *Capsella bursa-pastoris*

Shepherd's purse habit

Shepherd's purse triangular heart-shaped fruit

Shepherd's purse flowering habit and basal leaves

Mustard Family

SHEPHERD'S PURSE

Capsella bursa-pastoris

Description: Slender annual European weed with a basal rosette like a dandelion (without the milky sap), and clasping stem leaves. Tiny white 4-petaled flowers in dense terminal clusters. Heart-shaped, flattened seeds on elongated clusters. Common in disturbed, waste or cultivated sites from plains to subalpine.

Medicinal uses: Of all the mustards, this is the most used in herbal medicine with a long history of use to treat internal and external bleeding, diarrhea and inflammations with bleeding of the genitourinary, gastro-intestinal and upper respiratory tracts. It is also used as a mild diuretic, helping to remove uric acid from the body. It acts on the circulatory system to equalize blood pressure. The plant has proven uterine-contracting properties and has been used for premenstrual syndrome and as a childbirth aid. Externally, the plant is used for minor cuts and wounds, burns, nosebleeds, and as a mouthwash for mouth inflammations. Native peoples used the whole plant in tea for headaches, dysentery, stomach cramps, diarrhea and as a poison ivy wash. They used the seed pods to kill intestinal worms. The plant is a folk remedy for cancer. A homeopathic remedy is made from the fresh plant and is used in the treatment of nose bleeds and kidney stones.

Edible uses: The basal leaves in early spring are used like spinach, and have a peppery flavor. Once the flower stalk begins, I find the leaves too bitter to use. Blanching them (covering them with boards or canvas for a week or so) makes a more palatable plant. The fresh or dried roots have been used as a ginger substitute and can be candied by boiling in a sugar syrup. The seeds are tiny and somewhat peppery. The seeds and pods can be used to flavor soups and other dishes. Native Americans used the gray ash left after burning the plant as a salt substitute and tenderizer in certain dishes.

Notes: The seed has germicidal action against several micro-organisms. The mucilage of the seed, placed in water where mosquitoes breed in the spring, will kill the larvae and reduce mosquitoes in the area. One pound of seed destroys ten million larvae. Plants can be grown on salty or marshy land in order to reclaim it by absorbing the salt and 'sweetening' the soil. This species is a prime example of how a plant can be viewed as an annoying weed by some while for others it is cultivated for its wide range of uses, such as a cabbage-flavored spring green and an ingredient in a Japanese ceremonial rice and barley gruel.

Garden notes: The presence of shepherd's purse in our gardens indicates saline soils.

Caution: Avoid internal use during pregnancy. The seeds have been known to cause blistering of the skin.

Shepherd's Purse Could Be a Murderer!

There is some evidence that shepherd's purse seeds produce a sticky substance that, when moistened, can attract, entrap, kill and "digest" prey, making this a protocarnivorous (or paracarnivorous according to certain authors) plant (at least in this one stage of its life). This weedy species is typically found in low nutrient areas, and the small seeds have very little stored food sources. So these added nutrients from prey could be a huge benefit. Pretty neat idea, but not absolutely proven as of this writing.

John Barber, Tulane University, *Carnivorous Plant Newsletter*, 1978

Pennycress

Brassicaceae Family — *Thlaspi arvense*

Pennycress flower

Pennycress round heart-shaped fruit

Pennycress habit

Mustard Family

Pennycress
Thlaspi arvense

Other names: Field pennycress, fanweed, stinkweed

Description: Annual European weed 6 to 20 inches tall with most leaves clasping the stem. White, 4-petaled flowers in dense clusters elongate to bear flat, round heart-shaped pods with a large notch at the end. Common in dry disturbed or waste ground, fields and gardens from plains to subalpine.

Medicinal uses: The seeds and young shoots have been used as a treatment for the eyes. The seeds are considered tonic, anti-inflammatory, fever reducing, used to treat lung, kidney and reproductive system problems and rheumatic pain. The entire plant has historically been used to counteract poisoning. The plant is also anti-inflammatory, a blood tonic and blood purifier, diaphoretic, expectorant, fever reducer and hepatic. It has been used to treat various problems from sore throats, appendicitis, intestinal abscess and post-partum pain to endometriosis. It has broad antimicrobial activity against Staphylococci, Streptococci, Candida and others. Because of the high mustard oil content, it can be irritating to the system and should be used with caution. It is used in Tibetan and Traditional Chinese Medicine, but little used as medicine in the US.

Edible uses: Pennycress is a popular food plant in many parts of the world, especially Europe. The young, tender shoots are used like spinach, either raw or cooked as a potherb. It is too bitter for me, even after a couple of changes of water when cooking. But they can be mixed with milder greens to provide a nice punch. The dried seeds have a peppery, mustard flavor and can flavor soups, stews and other dishes. The seeds have been known to poison cattle, so users beware.

Notes: The seed contains semi-drying oil that is used for lighting and might be useful for making biodiesel (it is 36 to 40 percent oil by weight) and a nature-based weed killer.

> **Caution:** Use with caution since large doses can cause a decrease in white blood cells, nausea and dizziness. Fruits and seeds have been reported to be poisonous to livestock. Small amounts of the plant can taint cow's milk. Contact with pennycress may cause a reaction in people with sensitive skin.

Lamb's-Quarter Greens and Pennycress Wrap

These tiny little wraps are full of flavor.

Harvest several fresh (not dried out) pennycress seed pods. Also harvest several amaranth leaves, lamb's-quarters leaves, and the thin grass-like leaves of salsify. Place 1 amaranth leaf and 1 lamb's-quarters leaf on top of each other. Place a pennycress seed pod on top. Roll to form a tube or wrap. Tie together with the salsify leaf. Fill a small serving plate with the spicy bundles.

Variation: Add a bit of soft mild cheese to the wraps.

Foraging the Rocky Mountains— Liz Brown Morgan (page 101)

Brassicaceae Family — Mustards

Bittercress flower (above) and heart-shaped leaf (right circle)

Bittercress habit

Wild candytuft habit (left), clasping leaf and 4-petaled flower

Hoary Cress habit (below) and clasping, hairy leaf (right oval)

Mustard Family

MUSTARDS

WHITE FLOWERS

BITTERCRESS
Cardamine cordifolia

Other names: Heartleaf bittercress, large mountain bittercress

WILD CANDYTUFT
Noccaea fendleri ssp. *glauca*

Other names: Alpine pennycress

HOARY CRESS
Lepidium draba (Cardaria draba)

Other names: White top

Description: All mustards have four petals, alternate leaves clasping the stem and unique seed pods with 2 cells. Bittercress is a leafy-stalked perennial up to 24 inches tall with undivided, heart-shaped leaves. White flowers form a dense terminal cluster, producing slender, ascending pods. Found along streams from montane to subalpine. Blooms July to August. Wild candytuft is a slender perennial up to 8 inches tall with a dense tuft of white flowers. Stem leaves have a unique flap toward the base (an auricle), and the unique flattened seeds have a "beak" at the tip and sit horizontal along the stem. Early blooming and widespread from foothills up to the alpine. Hoary cress is a deep rooted non-native perennial up to 2 feet tall forming extensive colonies. White flowers are small and numerous atop the stem, leaves are grayish-green and lance-shaped. Considered an invasive weed of disturbed sites.

YELLOW FLOWERS

TANSY MUSTARD
Descurainia sophia

Other names: Flixweed, pinnate tansymustard, flaxweed tansymustard, herb sophia

WINTERCRESS
Barbarea orthocerus

Other names: American yellowrocket, erectpod wintercress

Description: Tansy mustard is an erect, lanky non-native annual up to 2 feet tall topped by bright yellow clusters of flowers in spring. The finely cut leaves are larger toward the base, and form spring rosettes. Fruits are slightly curved, linear pods. Often found in masses in disturbed areas. Wintercress can be annual, biennial or perennial with erect stems up to two feet tall. Pinnate leaves have a unique large, rounded leaflet at the end, and large clasping bases onto the stem. Bright yellow clusters sit atop the stem at first flowering, then elongate along the stem as the stem matures. Prefers moister disturbed sites, but can be found as a lankier specimen in drier disturbed areas as well. (Mustards are continued on the next page.)

Caution: Seeds of many mustard species, if consumed in large amounts may cause irritation of the stomach lining and may over-stimulate the thyroid. Prolonged contact (more than 10-15 minutes) may cause blistering on the skin when using mustards as plasters or poultices.

"Those who contemplate the beauty of the earth find reserves of strength that will endure as long as life lasts."

– Rachel Carson *The Sense of Wonder*

Brassicaceae Family — Mustards

Wintercress flower and seed pod (left) and habit (above)

Tansy mustard habit (right) and finely divided leaf (below)

Mustard Family

Mustards (continued)

Medicinal uses: The leaves and flowers of the wild mustards can be used as a digestive aid and carminative which help to reduce flatulence and gas. They are also high in vitamin C and historically used to treat scurvy. If the seeds are hot and pungent to the taste, they have medicinal properties and can be used externally in a plaster or poultice for various complaints or internally as a digestive stimulant and carminative. The seeds stimulate production of digestive juices in the stomach, thus aiding the digestive process. Externally the seeds are rubefacient, drawing blood to an area, thus increasing local heating and healing. There is history of use of the seeds in a lotion to treat frozen body parts, toothaches and sore throats. A traditional mustard plaster acts as a heating irritant and stimulant and is useful in treating paralysis, lung congestion, rheumatic and arthritic pains and stiffness. This is a centuries old decongestant treatment for bronchial congestion. The plaster is made from a ground seed paste sandwiched between pieces of cloth and applied hot to the affected area.

Edible uses: All the mustards are "edible", some definitely tastier than others. Because they are so easy to identify when in flower and are not look-alikes for much of anything else, experiment with the ones you like and how you prepare them. Typically the leaves are used when very young, becoming far too bitter as they mature. I have thrown flowers into my salads for a nice bite, and the seeds have varying flavors. Use the seeds as a flavoring in soups and stews and with meats. They are very strong in flavor. Seeds have been ground and mixed with salt and vinegar to make a sharp, peppery sauce (think Gray Poupon!).

Notes: Some species have been used for animal fodder.

Garden notes: The presence of many mustards in our garden often indicates hardpan or crusty surfaces, dry conditions, often a thin topsoil.

In Defense of Weeds

Richard Mabey, in his book *Weeds, In Defense of Nature's Most Unloved Plants* (page 289-291), says of weeds: "… it occurs to me that they are like a kind of immune system, organisms which move in to repair damaged tissue, in this case earth striped of its previous vegetation… Weeds opportunist lifestyles mean that their role-what they do-is to fill the empty spaces of the earth, to repair the vegetation shattered naturally for millions of years by landslide and flood and forest fire, and today degraded by aggressive farming and gross pollution. In doing so they stabilize the soil, conserve water loss, provide shelter for other plants and begin the process of succession to more complex and stable plant systems… Yet, once it became possible to attack them with machines and then chemicals, weeds slipped out of our understanding. Their appearance now sparks reflexes, not reasoning. They are regarded as inexplicable and impertinent intruders, quite unconnected with the way we live our lives. And in a radical shift of perspective we now blame the weeds, rather than ourselves. Every weed nuisance… has been the consequence of thoughtlessness and sometimes deliberate disruption of natural systems. Weeds are our most successful cultivated crop."

Calochortaceae Family   Mariposa lily   *Calochortus gunnisonii*

Mariposa lily flower and habit

Mariposa Lily

Calochortus gunnisonii

Other names: Sego lily, Gunnison's mariposa lily, cat's ear, star tulip, butterfly tulip

Description: Slender perennial from deeply buried onion-like bulbs. Stems are erect, typically unbranching except at the top, up to a foot tall with narrow, grass-like leaves. Flowers large, white to pink (older flowers), 3 petals with broad purple bands and gland-tipped hairs at the base of each petal. Found in dry meadows, sagebrush and aspen from montane up to alpine. Blooms from June to August.

Medical uses: There is some history of use of the whole plant tea during childbirth to facilitate afterbirth delivery. The juice of the leaves was applied to pimples.

Edible uses: As far as I am concerned, there should be no use of this beautiful, rare plant unless for an emergency. They have highly edible bulbs, tasting like potatoes or nuts eaten raw, boiled, roasted or baked. These bulbs are incredibly difficult to dig; being 5-6 inches deep with the stems breaking off right away. The leaves, seeds and flowers can be used, although they do not provide much bulk.

Notes: *Mariposa* is the Spanish word for butterfly, and *Calochortus* is Greek for "beautiful grass". The Shoshone word Sego means "edible bulb". *C. nuttallii* is Utah's state flower. The Ute people showed them to Mormon settlers, who had just lost their crops to Mormon crickets, saving many lives.

Garden notes: You can buy these bulbs from specialty bulb dealers. They are hard to find, but when you have a beautiful swath in full blooming glory in a dry meadow with butterflies swarming over them, won't you be happy you did?

Caution: When digging the bulbs, be absolutely sure of identification as they look very much like death camas.

How to Pick a Wildflower

There are many ways you can pick a wildflower.
You can lie on your stomach
in a meadow
and watch it grow.
You can stare at it
through a magnifying glass
to better discern each leaf.
You can paint it,
sketch it, etch it
in whole, or just in bloom.
You can take a picture then blow the picture up
larger, Larger, LARGER
and hang it on the wall.
Or you can pinch the stem between your fingers,
separate it from the Earth
and kill it.

— Carolyn Rickter

Harebells

Campanulaceae Family   *Campanula* spp.

Harebell habit

Harebell flower

Bellflower Family

Harebells

Campanula spp.

Common Harebell

Campanula rotundifolia

Other names: Mountain harebell, witch's thimble, lady's thimble, bellflower

Parry's Bellflower

Campanula parryi

Other names: Parry's harebell

Description: Both are delicate perennials, stem leaves alternate, linear and basal leaves rounded and toothed (frequently wilted by bloom time), flowers bell-shaped, purple, lavender to (rarely) white.

Common harebell is up to 24" tall, has several blue-lilac pendulous flowers per stem, pendulous seed capsules. Found in dry to moist, often rocky sites from montane to alpine. Blooms June to September.

Parry's bellflower is shorter at up to 12" tall, has typically one erect bright purple flower per stem, seed capsules erect. Lower stem leaves are fringed with white hairs. Found in moist meadows and aspen groves from montane to subalpine. Blooms July to August.

Medicinal uses: Native Americans chewed the root of **Common harebell** to treat heart and lung problems and used an infusion of the root as ear drops for sore ears. The whole plant was decocted as an internal and external treatment for sore eyes.

Parry's bellflower also had Native American uses. The root helped reduce inflammation. A poultice of the chewed or mashed root was applied to bruises. The dried plant has been used as a dusting powder for sores. The blossoms were chewed and the saliva applied to the skin to remove hair.

Edible uses: I tried the root once. They are extremely small, but very tasty, kind of sweet and nutty. Young leaves are edible as a potherb and the flowers can be added to salads.

Notes: Navajo wisdom says that if a woman eats harebells during her pregnancy, she will have a girl. They also rubbed Common harebell on their bodies for protection from injury while hunting and against witches. Collecting useful quantities of these delicate plants in the wild could decimate populations. Use them out of your garden instead.

Garden notes: Here is one flower you will love to have in your garden. They are adaptable to shade or full sun, will seed prolifically (if you let them) to form beautiful swales of color. I collect seed and sow it where I want it. I also found a beautiful white mutant on my friends land, and it is happily growing in my garden. They are readily available in nurseries.

Harebell

As summer fell
Away to autumn
I found
A solitary harebell
In the long grass
Where knapweed
And yellow rattle
Flower purple and gold.

That tiny splash
Of eggshell blue caught
The ripples of the wind
And danced,
And I danced too
With sheer delight.

— Val Moulton, 2012

Cannabaceae Family | Hops | *Humulus lupulus* var. *neomexicanus*

Hops habit and leaf and strobiles (inset)

Hops Family

Hops

Humulus lupulus var. *neomexicanus*

Other names: Wild hops, hopvine, common hop

Description: Vines trailing over the ground and rocks with opposite, palmately lobed leaves; both leaf and stem very rough to the touch. Flowers are unisexual; male flowers in loose racemes; females (also called strobiles) lack petals, but have obvious green, papery bracts surrounding them, turning yellow when dry (these are the "hops" used in brewing beer). Found in rocky areas and along fencerows of montane to subalpine. Note: The introduced cultivated hops are so close to our native plant that the uses are considered the same.

Medicinal uses: I really enjoy using this plant and it has a long history of worldwide medicinal use. The female flower heads, harvested in the fall, are used fresh or dried in teas, tinctures, washes, poultices and sleep pillows. They have pain relieving, nerve calming, sedative, antispasmodic, antiseptic, fever reducing and diuretic properties. They are very bitter and it is this bitter principle that stimulates the stomach, improves its tone, encourages the appetite and improves digestion. Hops are effective as a tea or tincture for mild cases of insomnia and indigestion, especially associated with worry and nervous exhaustion. A pillow of hops, placed inside a pillow case or under the head where the volatile oils can be inhaled, will have a soothing and sleep inducing action and promotes good dreams. Nursing mothers use hops to help increase their milk (which is why beer can have this same action). Hops are antimicrobial, especially against bacteria, both internally and externally. They are used to treat digestive infections from eating bad food. Poultices and washes of the strobiles can be used for inflammations, boils, bruises, cramps, swellings, skin ulcers, painful tumors, rheumatism and other external conditions. Whole plant extracts can be added to creams and lotion for skin softening affects. Internally they are also used to reduce fevers, coughs, diarrhea, worms, urinary stones, gas, cramping, Tuberculosis (TB), dysentery, uterine complaints and for symptom relief during menopause. This plant can be a valuable addition to any healing garden or natural medicine cabinet.

Edible uses: Use the female fruits (the hops) for brewing beer and ale. The young shoots are used as a potherb. Boil in water for 2 or 3 minutes, change the water and boil until tender. Some Native Americans used the seeds and male flowers in making bread. Some brewers have used the male flowers of the European species to make malt liquors with a strong bitter flavor.

Notes: The leaves and flower heads produce a brown dye. An essential oil from the strobiles is used in perfumery. The stems can be used for fiber, cloth and paper.

Garden notes: Hops grows easily and can be used to cover lattice for privacy or a fragrant, relaxing backyard arbor. They transplant easily from the wild. Typically the plants sold in nurseries are not our native variety. Hops can be somewhat aggressive, so care should be taken when choosing the site.

Caution: Skin contact with hops has caused dermatitis in sensitive persons.

Chenopodiaceae Family | Lamb's-quarters | *Chenopodium album*

Lamb's quarters stem markings at leaf axil

Lamb's quarters habit

Lamb's quarters flower

Lamb's quarters young plant

Goosefoot Family

LAMB'S-QUARTERS
Chenopodium album

Other names: Common pigweed, pigweed, goosefoot, netseed lambsquarters

Description: Alien annual of extremely variable forms from 1- 6 feet tall, very erect stems often striped with pink or purple. Leaves are long-stalked, alternate, simple and mealy white underneath; becoming diamond-shaped in maturity. Small, green, inconspicuous, ball-like flowers grow in dense, short spikes at the plant's tips, turning reddish-brown in seed. Common in waste and disturbed areas.

Medicinal uses: Today, if used as medicine, it is mainly for its nutrient value as a supportive food. It is mildly laxative. Historically, the leaves have been used as a wash or poultice for bug bites, sunstroke, rheumatic joints, swollen feet, inflamed eyes, headaches and dizziness. Seeds were chewed for urinary problems, the juice of the stems applied to freckles and sunburn and the root juice was used to treat bloody dysentery.

Edible uses: Steve Brill writes that the Chenopodium species that are odorless (such as lamb's quarters) are all edible. The resinous-smelling ones should only be used externally. Harvest the spring shoots less than 10 inches tall to use in salads and as a potherb (use just as you would spinach). Use the very young leaves all through the summer into the fall. The leaves can also be dried and powdered and placed in soups, quiches, with poultry… let your imagination go!! The leaves are super sources of vitamins, minerals and fiber. The shiny black seeds are also very nutritious. Collect in the fall and use in breads and muffins.

Notes: The young shoots produce a green dye. The crushed fresh roots have been used as a mild soap substitute.

Garden notes: The presence of lamb's quarters in our gardens often indicates a tilled or cultivated soil with high fertility or humus unless the plants are pale and stunted, then the fertility is low.

Caution: The plant contains saponins and oxalic acid. Adding calcium rich foods to the cooking process will counteract the oxalates. Consumption of large quantities is not advised but can safely be added to salads and cooked greens in small amounts. Some reports of sun sensitivity exist from folks eating large amounts of raw leaves.

Lamb's Quarters with Basil

2 lbs Lamb's Quarters
3 green onions, chopped
1/2 cup fresh basil, chopped
2 Tbs. butter

In a large saucepan add 1/3 cup water to lamb's quarters and onions. Cover and steam for 10 minutes. Drain. Add basil and butter. Toss to mix. Serve at once. Serves 6

— *The Rocky Mountain Foods Cookbook*, page 110

Convallariaceae Family — Solomon's seals — *Maianthemum* spp.

Starry false Solomon's seal flower

Starry false Solomon's seal habit

False Solomon's seal young shoots

False Solomon's seal green and ripe red fruit

False Solomon's seal flower

False Solomon's seal habit

Mayflower Family

Solomon's Seals
Maianthemum spp.

False Solomon's Seal
Maianthemum racemosum ssp. *amplexicaule* (*Smilacina amplexicaulis*)

Other names: Claspleaf Solomon's plume, many-flowered false Solomon seal, feathery false lily of the valley

Starry False Solomon's Seal
Maianthemum stellatum (*Smilacina stellatum*)

Other names: Star Solomon's plume, few-flowered false Solomon's seal, star-flowered Solomon's seal, little false Solomon's seal, star-flowered Solomon's seal, starry false lily of the valley

Description: False Solomon's seal is up to 30 inches tall with creeping rootstalks from which solitary arching stems arise. Leaves are alternate, broadly oval with prominent veins, and partially clasping the stem. White flowers, star-shaped with 6 tepals, in a pyramidal panicle. Berries green, turning splotched, then red. Found in rich soil, shady sites from montane to subalpine. Blooms May to June.

Wild lily-of-the-valley is similar to false Solomon's seal except smaller (to 16 inches tall) with narrower leaves, fewer flowers in a raceme with larger individual flowers. Found in similar sites, typically in the lower elevations. Blooms May to June.

Medicinal uses: These plants have a history of use treating stomach problems, colds, sore throats, internal pains, menstrual complaints and as a contraceptive. The roots were poulticed for sprains, boils, swellings, arthritis and rheumatism. The root is demulcent and expectorant for softening and moving out mucus in lung infections and can be used in a tea or a honey-based cough syrup. The berries are high in vitamin C and were used to prevent scurvy and rickets.

Edible uses: I have tried the very young shoots of false Solomon's seal just as they are emerging, boiled, and they taste like bitter asparagus. They can be confused with other plants at that stage, so only use them if you have positive identification. The rootstalks have been used by Native Americans, but eaten only after soaking them overnight in lye to remove the bitterness. The berries taste like peas when they are green, then turn sweeter when they ripen to red.

Garden notes: For a rich, shady site where a good ground cover is needed, bring in the Solomon's seals. They are hardy, cover a lot of ground, and their leaves are gorgeous.

> **Caution:** Do not consume any parts of this plant in large quantities as it tends to have laxative effects. Younger shoots can resemble poisonous cornhusk lily (*Veratrum tenuipetalum*).

Why False?
The name "false" comes from the plant's resemblance to the appearance and habitat of the eastern Solomon's seal (*Polygonatum* spp.). People gave many plants in the West the name "false" because they resembled plants from their previous homes. Often these plants are not related in any way. Hence, the danger of common names!

Crassulaceae Family | Stonecrop | *Sedum lanceolatum* var. *lanceolatum*

Stonecrop habit flowering and leaves (inset above)

Stonecrop in spring

Stonecrop Family

STONECROP

Sedum lanceolatum var. *lanceolatum*
(*Amerosedum lanceolatum*)

Other names: Yellow stonecrop, orpine, spearleaf stonecrop, narrow-leaf stonecrop

Description: A succulent perennial forming loose clumps, not strongly rhizomatous, up to 8" tall. Often red-tinged leaves cylindrical, several times longer than wide, alternate and sparse along flowering stem, otherwise forming tight basal rosettes. Bright yellow, dense clusters of 4-5-petaled star-shaped flowers. Found in dry, rocky sites from foothills to alpine. Blooms June to August.

Medicinal uses: The juice of this plant is high in vitamin C, slightly astringent and mucilaginous. A soothing poultice of the leaves or just the leaf juice can be applied to skin problems, wounds, burns and insect bites. As a tea taken internally, it is helpful for decreasing diarrhea, calming sore throats and treating lung problems.

Edible uses: I consider these a trail munchie, taking small nibbles as I come upon them. Try the flowers for a sweet treat. I find that most plants tend to keep their leaves all winter, so they could be considered a survival food.

Notes: An essence of stonecrop is said to "awaken and enhance illumination, bringing a moment of grace to those in the midst of struggle or despondency" according to Edward Bach. Sedum is Latin for "to sit" referring to how the sedum "sit" close to the ground.

Garden notes: I constantly am adding stonecrop to the stone walls and dry areas of my garden. As soon as a plant gets large enough, I will dig a small portion and place it in the rocky crevice, making sure to water it in well until it is well established. Once a good root system is set, you will not have to water them ever again; they might go dormant and look dead, but they almost always come back. The biggest mistake people make is overwatering them and killing them with kindness.

Caution: There have been reports of nausea after eating large amounts.

Drought Adaptations

Plants have adapted to drought situations for thousands of years. It seems unfathomable that they can thrive in sizzling heat, mere inches of rain and nasty soils. Water stress has created some amazing plant adaptions over time. Storage organs as modified root structures, trunk structures, stems, and leaves ensure water availability even in high drought times. Reducing the surface area of tissues through smaller leaves or tighter rosettes, and fewer branches lowers the rate of water loss. Dense hairs, such as those found on mullein, sagebrush, pussytoes and pearly everlasting can reduce water lost to wind and sun, as well as possibly reducing the temperature of the leaf. Waxes and glossy coatings are reflective as well as help reduce water loss. Many sedums use CAM photosynthesis instead of the usual C3 photosynthesis. Simply put, these plants open their stomata (the pores on the leaf surface allowing gases to pass in and out of a leaf) at night instead of during the day. Avoiding opening their stomata during the hottest part of the day reduces water loss.

Alfalfa

Fabaceae Family *Medicago sativa*

Alfalfa leaves

Alfalfa flower and leaves (left and oval above)
Alfalfa habit (below)

Pea Family

Alfalfa
Medicago sativa

Other names: Medic, lucerne

Description: An alien perennial up to 3 feet tall in large dense clumps. Flowers are dark purple to white, pea-like, with small leaves pinnately trifoliate compound. Found in cultivated areas, along roads and waste sites in foothills to montane.

Medicinal uses: Alfalfa is packed with nutrition and considered a whole body tonic used world-wide in herbal medicine for over 1500 years. The leaves and flowers, fresh or dried, can be used as a nutritive tonic that is stimulating to the appetite and can promote weight gain when needed. It is more a food or dietary aid than a medicine. The high mineral content, especially highly absorbable calcium and trace minerals, make it an excellent recuperative aid, especially when recovering from inflammatory illness, surgery and extreme stress, and mixes well with nettles and red clover for this use. It is also a supportive tea for chronic illnesses like arthritis, colitis, ulcers, anemia, asthma, diabetes, gastrointestinal disorders, etc. I prefer the fresher, greener taste of wild alfalfa but it still tastes a lot like grass, so is better mixed with other herbs like mint. It is also high in chlorophyll, flavonoids, digestive enzymes and has antibacterial properties. I include this herb in the nutritive tea I make for regular use combined with nettles, red raspberry, red clover, dandelion, horsetail and mint and with any other herbs that might strike my fancy that day.

> "Every flower is a soul blossoming in nature."
>
> — Gerard de Nerva

Edible uses: The younger leaves can be added to salads, but I find them very bland. They are used mostly in teas. Alfalfa also contains dozens of amino acids, making the plant high in protein. I am sure everyone is familiar with alfalfa sprouts we can buy in stores.

Notes: A yellow dye is obtained from the seed. The fiber of the plant has been used in making paper. The seed yields a drying oil used in paints, varnish, etc. Alfalfa is a potentially excellent source of biomass. It is a major feed crop for domestic animals and a component of hay.

Garden notes: Plants are very deep rooting, and are able to fix large quantities of atmospheric nitrogen; this makes them one of the very best green manures. The plant can be grown as a low dividing hedge in the vegetable garden.

Caution: Generally alfalfa is considered safe and nutritious. There are, however, some contraindications and side effects to consider for internal use of alfalfa plants and sprouts: contraindicated during pregnancy and lactation, avoid for people with hormone sensitive cancer, avoid for people with gout (due to purines), not recommended for persons with autoimmune disease or taking anticoagulant medications. The constituent canavanine can cause the recurrence of systemic lupus erythematosus (an ulcerous disease of the skin) in patients where the disease had become dormant. Regular long term use could trigger photosensitivity and red blood cell deficiency. Collect only in chemical free areas.

Red clover

Fabaceae Family — *Trifolium pratense*

Red clover habit

Red clover trifoliate leaf

Red clover flower and leaf with distinctive v-shaped chevron design

144

Pea Family

Red Clover

Trifolium pratense

Description: A non-native perennial from Europe growing up to a foot tall with a ball of red-purple to pink pea-like flowers at each stem end with 1 to 2 short-stalked leaves at their base. Trifoliate leaves with a prominent cream-colored "v" (called a chevron) arise from flower stalks which can be erect or spreading. Blooms most of the summer. Found in lawns, meadows and disturbed sites.

Medicinal uses: Red clover is safe and effective with a long history of use worldwide. Native people used the leaves and purple flowers of this common plant for treating fevers, whooping cough, menopause, cancer and as a blood cleanser. The Eclectics recommended red clover for cough, bronchitis, and TB, but more importantly as a cancer treatment. Recent studies support its anti-tumor actions. During the late 19th and early 20th centuries, red clover was the major ingredient in many patent medicines. Often called the "prize" herb valued for its alkalinizing effects. Red clover flowers are used internally for skin complaints, especially eczema and psoriasis. They may be used with complete safety in cases of childhood eczema, all types of cancers, fibroids, cysts, chronic degenerative diseases, gout, whooping cough, asthma, dry cough and to improve appetite and build blood. The flavonoids in the flowers are estrogenic and are useful for menopausal complaints. Externally, poultices of the herb have been used for local applications to cancerous growths. A homeopathic remedy is also made from red clover.

Edible uses: I personally find the white clover (*T. repens*) flower sweeter than the red clover, but the red clover is a richer flavor. Try all the clovers and discover your favorite. The flowers are a favorite sweet trail munchie, and are added to salads and drinks. They can be dried and ground into an excellent flour. The leaves are best in early spring when they are very tender. Many people have difficulty digesting the leaves so use them in moderation; boiling or soaking them in salt water for several hours can make them more digestible, but much of the nutrition is destroyed.

Notes: The flowers produce a yellow dye. Red clover is the state flower of Vermont.

Garden notes: A good green manure with nitrogen fixing properties. It is an important feed plant for the caterpillars for many butterfly and moth species and bees like it, too. White clover is grown with grass mixtures for land reclamation.

Caution: Hemophiliacs or those with 'thin' blood should not use red clover regularly. Because it has some estrogenic effects, it should not be consumed in cases of estrogen receptor-positive tumors. Only use fresh plants parts, those turning brown can have toxic alkaloids.

Red Clover Through the Ages

Red clover is one of the world's oldest agricultural crops, cultivated as forage since prehistoric times. Because of its importance in early agriculture, red clover has a long history as a religious symbol. The ancient Greeks, Romans, and the Celts of pre-Christian Ireland all revered it. Early Christians linked the plant to the Trinity, and some say red clover is the model for Ireland's symbol, the shamrock. Red clover was also the model for the suit of clubs in playing cards. During the Middle Ages, red clover was considered a charm against witchcraft.

— *Herbalpedia*

Gentians

Gentianaceae Family — *Gentianopsis thermalis & Gentiana parryi*

Bottle gentian flower

Fringed gentian flower

Caution: Pregnant women and those with digestive problems or high blood pressure should avoid internal use. Excess amounts can cause nausea and stomach disturbances.

Gentian Family

GENTIANS
FRINGED GENTIAN

Gentianopsis thermalis

Other names: Rocky Mountain fringed gentian

BOTTLE GENTIAN

Gentiana parryi
(Pneumonanthe parryi)

Other names: Parry's gentian

Description: For ease of identification we will cover only the "shorter" gentians here. See Green gentian for additional information. Our local gentians are typically higher elevation plants, under 12" tall, flowers are intense blue to rose or purple, with one white species, tubular and tubular-funnel shaped. Most flowers are responsive to light, opening only in bright sunshine and closing in shade (even the shade of your hand after several minutes). All bloom in late summer. Leaves are opposite, typically joined around the stem at the base. Here are a couple of our more common gentians:

Fringed gentian Gentianopsis thermalis is a slender annual with large, deep violet flowers twisted together at the top with a "fringe" on the petals. Found en masse in wet meadows from upper montane to subalpine.

Bottle gentian Gentiana parryi is a showy perennial with relatively large blue-violet barrel-shaped flowers bearing small teeth between the petal lobes in tight terminal clusters. The most common large gentian of upper montane and subalpine meadows.

Medicinal uses: Gentian roots are considered one of the best digestive tonic bitters with a long history of uses for the digestive and urinary systems. I rarely find enough of these beautiful plants to justify harvesting. I substitute green gentian which has similar uses but is a little more stimulating.

Edible uses: Enzian (gentian) schnapps is a "bitters" found throughout Europe, made from the roots of the yellow gentian of the Alps. Were the blue gentians used? I could not find any notes on this.

Garden notes: Oh, to have these beauties gracing your garden, what a joy!! Our native species can be hard to find, but any of the low growing species can be a boon to bees, so give them a try. They can be a little demanding as far as growing conditions, but try a variety of sites and soils until you find the right combination.

> ### God Made a Little Gentian
>
> God made a little Gentian—
> It tried—to be a Rose—
> And failed—and all the Summer laughed—
> But just before the Snows
>
> There rose a Purple Creature—
> That ravished all the Hill—
> And Summer hid her Forehead—
> And Mockery—was still—
>
> The Frosts were her condition—
> The Tyrian would not come
> Until the North—invoke it—
> Creator—Shall I—bloom?
>
> — Emily Dickinson

Green Gentian

Gentianaceae Family — *Frasera speciosa*

Green gentian flower

Green gentian habit blooming stalk (above left)

Green gentian rosette habit

Gentian Family

GREEN GENTIAN

Frasera speciosa (Swertia radiata)

Other names: Monument plant, elkweed, deer tongue, deer's ear, Cebadilla

Description: A monocarpic perennial, meaning the plant lives for up to 80 years as a huge, leafy rosette, then sends up a tall flower stalk, sets seed and dies leaving only seeds to carry on after their long life. Rosette leaves are bright green, smooth and up to 12 inches long arising from a very large taproot. Flowering stalk up to 6 feet tall, creamy greenish colored flowers open star-shaped with four petals, purplish markings and fringes of long hairs near the center. Masses of flowers cover the stalk. Found in moderately dry areas from montane into alpine (where they are very short). Blooms June to August.

Medicinal uses: Green gentian root is very bitter. Digestive bitters have been used for centuries to improve digestion by stimulating appetite and digestion. Digestion starts in the mouth and the bitter taste stimulates saliva which in-turn starts the cascade secretion of other digestive juices along the intestinal tract. Bitters are taken 15 to 20 minutes before a meal. A brandy extract of the root is frequently used or a pinch of powder taken in water. Small amounts are plenty and larger amounts can be laxative and irritating to the large intestine. This root also has fungicidal properties used as a powder for athlete's foot, jock itch, etc. A root tincture can be used externally for ringworm and as an emergency first aid antimicrobial. A root salve can be used for killing lice and scabies.

Edible uses: The fleshy root can be eaten raw, roasted or boiled. It is usually mixed with other roots or leafy greens. I do not like to dig this amazing plant, so have tried it only once. That was good enough for me.

Notes: A valued pollinator plant.

Caution: Pregnant women and those with digestive problems should avoid internal use. Excess amounts can cause nausea and vomiting and other stomach disturbances. There is some indication that a short blood glucose drop following ingestion occurs in folks with fragile blood sugars.

Green Gentian Cycles of Blooming

In 2003 there was a virtual "forest" of green gentians covering the dry hillsides around the valley. Three years later I could hardly find one blooming. Green gentian spends up to 80 years as a vegetative rosette, then flowers, sets seed and dies. This is a monocarpic plant. Many years of research, up to 2010, by Dr. David Inouye at the Rocky Mountain Biological Laboratory in Gothic, Colorado showed that "A very wet July and August 4 years previous seems to be the cue for [Frasera speciosa] to start preforming a flower stalk". The flowering seems to be coordinated in a 2-4 year period where hundreds and thousands of plants burst into bloom. The flowering stalks begin to grow only microscopically for three years, then, suddenly, grow up to 6 feet tall in a month. Remember, these are plants that can live to 80 years, so how can they ensure cross pollination? This coordination of bloom time where there are peak years in a population (up to 90% of the mature plants blooming) and low years increases the odds of cross pollination. The last epic bloom period was 2010, with 2013 being well above average.

Geraniaceae Family — Geranium
Geranium viscosissimum & *G. richardsonii*

Sticky geranium habit (above left) and flower (top right inset)

Sticky geranium fruit

Richardson's habit and flower (inset)

Geranium Family

GERANIUMS
STICKY GERANIUM

Geranium viscosissimum

Other names: Sticky purple geranium, stork's bill, wild geranium, cranesbill

RICHARDSON'S GERANIUM

G. richardsonii

Other names: White geranium, cranesbill, stork's bill, wild geranium

Description: **Sticky geranium** is an attractive mounded perennial up to 2 feet tall. Leaves are mostly basal, deeply palmately lobed with 5-7 lobes. Flowers are pink to magenta, strongly purple-veined, with 5 petals, with several flowers on open clusters. Fruits look like a crane's bill. They open by drying and splitting lengthwise from the base and recoiling (like a catapult). Found in dry, open or shaded sites from foothills to lower subalpine. Blooms May to September.

Richardson's geranium is very similar to sticky geranium except the plants are slender, with thinner leaves and white petals. They are found in shadier habitats, but can hybridize with sticky geranium where they meet at meadows edge.

Medicinal uses: Many native plants have astringent properties. It is helpful to be able to identify a few of these because they can be great first aid remedies when out in the woods and faced with unexpected emergencies like bleeding, insect bites and diarrhea. The root and leaves of geraniums are strongly astringent with high tannin content. The root can be powdered for a poultice or decocted as a tea or wash to stop bleeding, decrease diarrhea, as an enema for bleeding hemorrhoids, as a gargle for sore throat and mouth inflammations and a douche for vaginitis. The tea may help with excess menstruation or postpartum bleeding. A slice of the fresh root applied directly on the painful area can be a first aid for gum or tooth infections. A poultice or infusion of the leaves is a topical remedy for insect bites, rashes and minor skin irritations. Native Americans used an infusion of powdered roots taken internally or powdered leaves used as snuff for nosebleeds.

Edible uses: The flowers and leaves are highly astringent and their texture is rather odd. Recommended only as an emergency food.

Notes: The dark veins on the petals are "landing strips" for insects capable of seeing the ultra-violet light the lines reflect, leading them to nectar at the base of the petals.

Garden notes: Sticky geranium is a fantastic plant for your landscape. Adaptable to very dry or to more moist conditions, these plants bloom almost all summer. They will bloom longer with some extra moisture, but will thrive with only Mother Nature's moisture. They are readily available in the landscape trade, or easily transplanted with young starts.

Watch Out for Catapulting Geranium Seeds!

This unique seed dispersal comes from the ingenious design of the "crane's bill" seed case. The five-seed casings are aligned in a long beak with the seeds at the base. As the seed case dries, tension is placed on the tissues of the "bill" which eventually tightens so much that it pulls the seed away from the base in a super fast motion, catapulting the seed into the air, up to 30 feet! After landing, this unique seed has a hygroscopic seed "tail", coiling when wet and straightening when dry. These actions can "move" the seed across the soil and even help push the seed into the soil.

Helleboraceae Family   Marsh-marigold   *Caltha leptosepala*

Marsh-marigold leaf and flower

Marsh-marigold habit

Hellebore Family

Marsh-Marigold

Caltha leptosepala
(Psychrophila leptosepala)

Other names: Mountain marsh-marigold, western marsh marigold, elkslip, elk's lip marsh marigold, white marsh-marigold, slendersepal marshmarigold

Description: Fleshy perennial herb up to 6 inches tall (flowering stems) with mostly basal, fleshy, glossy oblong to heart-shaped leaves. Single flowers atop a leafless stem are white with a bluish tinge on the underside (buds tend to be bluish because of this) with 5 to15 sepals (no petals) and numerous stamens. Common in snowmelt basins and wet sites in subalpine and alpine. Blooms as snow is melting.

Medicinal uses: The whole dried plant is an antispasmodic and expectorant. It should be used in small doses, short term, to stimulate the flow of mucus in the respiratory system. A poultice of the mashed roots has been applied to inflamed wounds, insect bites and bruises.

Edible uses: The buds are eaten raw and have a pleasant flavor; although in today's environment of highly questionable water practically anywhere, I would hesitate eating any parts raw unless it is an emergency. Leaves have been recommended both raw and cooked (again, I would cook them). Younger leaves are definitely better as they are tough and somewhat bitter when older. The fleshy roots cooked resemble sauerkraut. It would be best to keep marsh-marigold only as an emergency food. The flower buds of the yellow eastern species (*Caltha palustris*) are pickled and used as capers.

Garden notes: Create a bog garden where water from your roof collects and sow the seed collected from marsh-marigold and enjoy! The bright yellow *C. palustris* is common in the nursery trade and is a gorgeous early summer addition in the bog garden.

Caution: The whole plant, especially older parts, contains the poisonous glycoside protoanemonin which is dissipated with heating or drying. The sap can irritate sensitive skin and using gloves is recommended.

"In joy or sadness, flowers are our constant friends. We eat, drink, sing, dance, and flirt with them. We wed and christen with flowers. We dare not die without them. We have worshipped with the lily, we have meditated with the lotus, we have charged in battle array with the rose and the chrysanthemum. We have even attempted to speak in the language of flowers. How could we live without them? It frightens one to conceive of a world bereft of their presence. What solace do they not bring to the bedside of the sick, what a light of bliss to the darkness of weary spirits? Their serene tenderness restores to us our waning confidence in the universe even as the intent gaze of a beautiful child recalls our lost hopes. When we are laid low in the dust it is they who linger in sorrow over our graves."

— Kakuzo Okakura *The Book of Tea*

Lamiaceae Family   Giant hyssop   *Agastache urticifolia*

Giant hyssop habit

Giant hyssop flower (above left oval)

Giant hyssop immature leaf

Mint Family

Giant Hyssop
Agastache urticifolia

Other names: Nettle-leaf horsemint, nettle-leaved giant hyssop

Description: Perennial up to 2 to 2½ feet tall from branching woody root crowns. Leaves are opposite, heart to egg-shaped and coarsely toothed, on a square stem. The rose to purplish flowers are in a dense terminal spike; stamens are exerted, producing a fuzzy look to the flower. Found in rich soils of aspen and open forests from foothills to subalpine, generally in dense colonies. Blooms June to August.

Medicinal uses: This member of the Mint Family has similar properties to all mints but without the familiar mint taste. The leaf and flower tea is used for indigestion, colds, coughs, chest pain associated with coughing, fevers, intestinal gas, and rheumatism. Some Native people used the tea to strengthen a weak heart. Externally it was used as a burn dressing, to lower fevers and mixed with other herbs for poison ivy itching.

Edible uses: The quality of flavor from plant to plant seems to vary quite a bit. Some have a sweet anise flavor, others taste like dirty socks. Once I have found a patch that I like, I go back to it year after year. The seeds are eaten raw or cooked, leaves flavor soups, stews and other hot dishes. Steep the leaves in hot water (brew weakly) for a comforting tea.

Garden notes: These are easy plants to introduce into a landscape of richer soils and some moisture. They transplant easily from a portion of root with several shoots, or seed. They can become aggressive.

Doctrine of Signatures

The Doctrine of Signatures began as a spiritual philosophy that held that God had marked everything created with a sign. It was popularized by Paracelsus in the 1500s, used in China and among Native Americans and continued to be popular until the early 1900's. It states that a plant looks like what it is good for; that a plant's color, shape, taste, smell, etc. can indicate its use and body part affinity. Some examples are: yellow, gold and orange flowers and roots are often kidney (yellow urine) and liver (yellow bile) and digestive remedies. Red flowers or fruits indicate actions on the blood and heart: hawthorn berries are a heart tonic, red clover flowers are blood cleansers and nourishers. Blue to violet flowers correspond to the higher chakras and indicate actions on the head, nervous systems and emotions: blue skullcap and giant hyssop have been used for headaches. Smells can indicate actions: a dirty socks smell, like found in valerian root, is often sedative and anti-spasmodic and really bad smelling plants like hemlocks are often poisonous in large amounts. Though some scientific types like to discredit it, I and many of my colleagues have found it to be surprisingly accurate and it continues to be popular today among folk herbalists. This method uses intuition and encourages a direct experience of learning Mother Nature's plant language. I don't rely solely on this method for figuring out the action of a plant, but it is a fun way to "meet" a plant, then do the research and see if the "signature" holds true.

Lamiaceae Family — Field mint — *Mentha arvensis*

Field mint flowers at leaf axils and square stem

Field mint habit

Field mint flower and leaf

Mint Family

Field Mint
Mentha arvensis

Other Names: Poleo, corn mint, wild mint, Indian mint

Description: Spreading perennials up to 18" tall with square stems, lance-shaped leaves, sharply toothed, opposite the stem. Highly aromatic foliage. Lavender flowers in compact whorls from middle to upper leaf axils. Found in wet areas, open or shaded sites, from foothills to montane. Blooms July to September.

Medicinal uses: Like its cousin, peppermint, field mint contains menthol and is an old valuable remedy for upset stomachs. It is useful for colic, indigestion, expelling gas and relieving spasms of the digestive tract. Good sipped as an after dinner tea when digestion seems a bit off. The leaf tea or tincture can also be used to treat colds, coughs, headaches and fevers. This plant also contains pulegone which can be used to stimulate scanty or delayed menstruation, especially if accompanied by bloating or painful cramps. It is a lovely strong mint flavoring for tea blends. I love this plant!

Edible uses: Wonderful, refreshing mints! The stems and leaves are used in teas, cold drinks, and even wine. The plants can be eaten raw as greens, but they are usually cooked with soups, stews or other meats, or used to flavor jams, jellies and candies. I put a fresh leaf into each ice cube partition for a nice touch to iced tea. The leaves can be sugared for cake toppings or sides for desserts. Dry the leaves and use them powdered over meats, in fruit juices, yogurt and fruit salad.

Notes: The Native Americans often stored dried meats with powdered mint because rats and mice have a strong dislike for mint. Whole plants were hung in dwellings as air fresheners, and rubbed on the body as perfume. Dried, crushed leaves placed along ants' entry paths will deter them from coming into your house.

Garden notes: Field mint is best in moist, rich areas, and is a valued bee plant. In fact all the Mint Family members are fantastic pollinators, great to deter deer and other browsers and insect pests.

Caution: Not for use during pregnancy. Large doses, especially of the essential oil, can cause abortion. Use caution when giving infants or young children food or medicines containing menthol or peppermint as the strong fragrance could gag or choke them. Be sure to thoroughly wash plants pulled out of streams and other waterways.

Wild Mint-Lemon Cooler

2 lemons
12 sprigs Wild Mint
1/3 cup granulated sugar
4 cups cold water

Squeeze juice from lemon; add Wild Mint and sugar. Mash with potato masher to extract flavor from mint. Let mixture set for 30 minutes. Stir in water. Pour over cracked ice and serve. Serves 2

— *The Rocky Mountain Wild Foods Cookbook* by Darcy Williamson (page 212)

Lamiaceae Family — Bee balm — *Monarda fistulosa*

Bee balm flower shows tubular petals that hummingbirds love

Bee balm flower and opposite leaves on square stem (left)

Bee balm habit

Mint Family

Bee Balm

Monarda fistulosa

Other Names: Horsemint, Oswego tea, wild oregano, wild bergamot, mintleaf beebalm

Description: Highly aromatic perennials with 4-sided leafy stems up to 2 feet tall from spreading rhizomes. Leaves opposite, lance to egg-shaped. Flowers 2-lipped, upper lip long and arching over lower lip; bright rose to purplish terminal clusters above several leafy bracts. Found in moist to moderately dry open to open-forested sites from foothills to montane. Blooms June to August.

Medicinal uses: The leaf and flower tea is an effective diaphoretic and infection fighter, making one "sweat out" infections which is an excellent way to treat fevers, colds and flu. The leaves contain an antiseptic and disinfectant compound that is used in some mouthwashes. The tea can also be used as a mouthwash, gargle for throat infections, and to expel intestinal gas. This anti-inflammatory, anti-infective tea can also be used to treat a wide variety of other ailments including colic, insomnia, kidney and respiratory problems, chronic urinary tract and vaginal yeast infections, nosebleeds, chest pain, headaches, suppressed menstruation, cramps, indigestion and worms. The inhaled steam helps fight lung inflammation and decrease the spread of infections. Externally, its antiseptic and antimicrobial properties make it useful as a poultice or wash for wounds, skin eruptions, burns, and a wash for sore eyes. I love the beauty, smell and power of this plant ally.

Edible uses: The taste of bee balm is more like oregano than mints, so they can be used accordingly to flavor stews, soups and meat dishes. One of my favorite teas is a mild green tea with horsemint leaves (younger ones are better) and honey. Earl Gray tea with oil-of-bergamot is extracted from the tropical tree, orange bergamot (Citrus aurantium). Our native bee balm is a little bitterer than this, but has the essence of this tree. Drying can remove some of the bitterness.

Notes: Native Americans would take the dried leaves and sprinkle them over meat and drying berries to repel insects.

Garden notes: Being a mint, I thought this would be a super easy plant in my garden. I have tried it in a richer soil area under aspen, a west exposure with native soil, a south exposure with rich soil, and a south exposure with a little shade from shrubs in native soil. The most spectacular plant I have found in the wild was growing in gravel alongside a road facing south. It had seeded in there and was HUGE! It totally eroded all my images of bee balm in nature in rich soil under aspens or shrubs. This is a valuable pollinator plant, and hummingbirds love it.

Bee Balm Honey

In a glass jar, cover fresh bee balm flowers with raw honey (raw honey is healing and antimicrobial). Infuse for 4 to 6 weeks in a warm place, strain and store in a cool, dark place. Keep on hand as a topical treatment for burns and wounds. Internally, it is good for the sore throats, coughs, congestion and fever of colds and flu. Use like a syrup and add to tea or take it straight.

Self-heal

Lamiaceae Family — *Prunella vulgaris*

Self-heal flower

Self-heal habit

Mint Family

Self-Heal

Prunella vulgaris

Other names: All heal, heal-all, woundwort, common self-heal.

Description: A somewhat prostrate perennial, growing up to 12" tall with the typical Mint Family square stem and opposite leaves which are oblong or egg-shaped, narrower but not smaller upwards on the stem. Unlike other mints, self-heal has a dense terminal spike of purple flowers, and has very little aroma. Size varies with the climate, soil quality and moisture. Found in many moist disturbed sites (lawns included), but thrives in rich, moist soils from lower subalpine to foothills. Blooms May to August.

Medicinal uses: As you can tell by the various names, this plant has been used for just about everything. I love this plant! Some years it is very plentiful, other years a surprising find. The whole above ground plant dried, tinctured, juiced or infused in oil for salves is used for its soothing, healing, astringent, antispasmodic, antiseptic, as well as many more properties. Internally and externally it is a gentle remedy for everything from colds, fevers and sore throats to wounds, upset stomachs and headaches. Get to know this one and you will have a wonderful healing ally.

Edible uses: The entire plant is edible, especially when young and tender, both raw and cooked. The leaves make a pleasant tea. Unlike the other mints, self-heal is fairly bland.

Notes: Self-heal is a favorite of bees and butterflies. Olive-green dye can be obtained from the leaf and stem.

Garden notes: There is a plethora of self-heal in the garden trade. They are stunningly beautiful and extremely tough ground covers of dry (although they prefer more moisture) shade areas (some very aggressive!). As far as whether they have retained the valuable constituents of our native species, I do not know. Establishing the native in your garden is as simple as taking a cutting and watering it until roots are well formed.

Visual Cues for Pollinators

One of the most fascinating cues to pollinators are the visual cues, especially colors. The range of colors that our human eyes "see" is different than what most insects perceive. Red, at the lowest frequency, and violet, at the highest frequency, are the rainbow of light we see. Honeybees, bumblebees and many diurnal butterflies have true color vision but the red-end vision is very poor. They make up for this by seeing colors well into what we call "ultra violet", or UV light. What would this "look" like? We can only guess, but we have tried our hand at capturing them using specialized photography. These photos show that our yellow Potentilla anserina is not just yellow, but has a striking band of color near the center of the flower that only insects can perceive. Our Geranium viscosissimum has striking lines leading to the center of the flower; right to the nectaries and pollen. It seems that flowers have developed morphology in tune with insect vision (not OUR vision…gasp!) and created guides for insect landing and nectar finding.

One of the most amazing websites to see flowers in all of insect view glory is: http://www.naturfotograf.com/uvstart.html

Liliaceae Family | Glacier lily | *Erythronium grandiflorum*

Glacier lily fruit

Glacier lily flower

Glacier lily habit

Lily Family

Glacier Lily
Erythronium grandiflorum

Other names: Yellow fawn lily, snow lily, trout lily, adder's tongue, dog-toothed violet, avalanche lily

Description: An early spring perennial emerging from deep bulbs with 2 large, strap-shaped glossy leaves, and bright yellow flowers of 6 recurved tepals and exerted stamens. Generally one to two flowers per stem, but can be many (13 have been counted). Found in rich-soiled rocky areas and moist aspen forests from montane to alpine. Bloom May to June, following the snow line.

Medicinal uses: There is a history of use of the crushed root poultice for boils and skin sores. The leaf tea has antibacterial properties and can be used as an antiseptic wash for cuts, scrapes and sores. I personally refrain from using glacier lilies medicinally because of the threat of overharvesting.

Edible uses: Historically, the bulbs were collected in large quantities, eaten raw or dried. Strings of dried bulbs were a popular trade item. These dried bulbs were soaked, then boiled or steamed. Once cooked, the bulbs became a rich chocolate brown color and were quite sweet. Today the bulbs should not be collected except in extreme emergency because, even though they can be locally abundant, they are a beautiful lily taking many years till flowering stage and in many areas of the West have been overharvested. Instead of digging the bulbs, place the tasty flowers, leaves or green seed pods in salads, or put the green seed pods in stir fries. Be very frugal at first, as these are an emetic.

Garden notes: I keep trying to introduce these to my garden, with limited success. Every summer I collect a handful of seeds and sow them in different areas to try and get them established. The seedlings only produce one leaf, and are very small, and it takes several years for the corm to build enough reserves to push up two leaves, and still more years before a flower will emerge. For this beauty, I can be very patient!

> **Caution:** Eat this plant in moderation as it can be a strong emetic. Collect the bulbs when the plant is flowering to ensure positive identification.

A Root is a Root

The typical true root consists of a primary root and secondary roots. There are many variations on the root, where a part of the root has been modified. Some of these modified roots found in our area include: **contractile roots** pull bulbs or corms (and rarely taproots) deeper in the soil by expanding radially and contracting longitudinally (glacier lily and dandelion); **haustorial roots** are roots of parasitic plants that absorb water and nutrients from another plant (mistletoe); **propagative roots** form adventitious buds that develop into aboveground shoots, termed suckers, which form new plants (Canada thistle, aspen, cattails, nettles); and **storage roots** include some taproots and tuberous roots and are modified for storage of food or water (salsify, yampah). There are many other types of roots, but these are the most commonly encountered in our area.

Linaceae Family — Wild blue flax — *Linum lewisii* var. *lewisii*

Wild blue flax flower and fruit

Wild blue flax habit

Flax Family

Wild Blue Flax

Linum lewisii var. *lewisii*
(*Adenolinum lewisii*)

Other names: Western blue flax, blue flax, perennial flax

Description: A perennial up to 2 feet tall in small clumps of many slender stems with grayish linear leaves along the stem. Pale to intense 5-petaled blue flowers top the stem, opening singly in few-flowered clusters and fading rapidly. Seeds are brown capsules with several shiny brown-black seeds. Found in montane to lower subalpine in drier, poor soil sites. Blooms most of the summer.

Medicinal uses: This plant was used by Native peoples for numerous problems. The whole plant steeped and used for upset stomach and gas, colds, coughs and rheumatic pains. The root was steeped for eye medicine and poultices of the fresh leaf for swellings, inflammations and eye problems. The seeds are emollient and used externally for sores, burns and boils. The oil from flax seeds contains essential fatty acids which lower blood fats and cholesterol, increase immune function and reduce clotting. The seeds are a source of fiber, mildly laxative, and soothing to the mucous membranes. The European relative of this plant (*L. usitatissimum*) is the variety most commonly commercially cultivated for these uses.

Edible uses: The shiny dark brown seeds are rich in oil and can be dried, roasted and ground into flour, or ground and eaten raw (in small quantities).

Notes: The long stems were slightly pounded and the strong fibers separated and used for ropes, thread and fiber. The whole plant infusion was used for a skin and hair wash.

Garden notes: Blue flax makes an excellent xeric garden plant. Their fine, airy texture can be used in masses to add color almost all summer. Seed is very successful, but plants can be planted. Let the plants always reseed as the plants are not reliable year after year.

Caution: Immature or raw seeds can cause toxic reactions. Livestock has been poisoned eating flax.

Fiber Plants

People have been using plant fibers for thousands of years to make clothing, paper, rope, etc. Cultivated flax (*Linum usitatissimum*) is one of the earliest known fiber plants. Flax linen was used to shroud Egyptian mummies. The stems of our species were used to make string and cords for baskets, mats, snowshoe mesh and fishing nets. The dogbanes were one of the most important fiber sources for making thread, twine, cordage, bags, fabric, paper, bowstrings, etc. The straight 2 to 3 year old stems were harvested in the fall as the seed pods were forming. Fiber from Indian hemp (*Apocynum cannabium*) is longer and stronger and preferred over our spreading dogbane (*A. androsaemifolium*). Stinging nettle produced good quality fiber for cloth and was often cultivated for this purpose. The stem fibers from hops produced a coarse cloth. Milkweed stems produced a tough fiber for cloth and was usually harvested from dried dead stems in the fall and winter. Ute basketry was produced from Sumac (*Rhus trilobata*) or willow. Native peoples often used fire every few years to manage plants for optimum fiber production.

Yellow pond lily

Nymphacaceae Family — *Nuphar polysepala*

Yellow pond lily habit

Yellow pond lily

Yellow pond lily dropping seeds

Water-Lily Family

Yellow Pond Lily

Nuphar polysepala (N. lutea, Nymphaea polysepala)

Other names: Spatterdock, Rocky Mountain yellow pond lily, Rocky Mountain cowlily, yellow waterlily, Wokas, nack-rose

Description: Aquatic perennial with fleshy stems up to 6 feet long from massive rootstalks. The leaves float on the water, are leathery and roundish. Yellow flowers sit just above the water, have 6-9 waxy petal-like sepals, with 10-20 tiny true petals hidden below numerous stamens and large yellow disc-like stigma. Found in shallow, quiet water from foothills to subalpine. Blooms June to August.

Medicinal uses: The roots have a history of use in teas to treat sore throats, tuberculosis, heart disease, inflamed gums, diarrhea, gonorrhea, rheumatism, blood diseases, impotence, stomach inflammation and gallstones. Externally mashed rootstocks have been used as poultices on wounds, swellings, swollen and inflamed joints.

Edible uses: The seeds and massive rootstalks were valuable food for many tribes. The dried capsules were broken open and the hard seeds were stored as they were for winter use, or they could be parched, or dried and ground into meal. Try taking the seeds and parch them in a frying pan over low heat. The seeds might pop open a little (like popcorn), which can be eaten, or then pounded or lightly ground and the hard shells removed by winnowing, shaking or picking them out. The kernels were ground into a meal with many uses. One way is to mix 2 cups boiling water with 1 cup meal, boil for 15 minutes stirring constantly, then simmer over very low heat for an hour. Eat like hot mush, or let it cool and slice and fry in butter. The rootstalks can be **extremely** bitter, but can be a valuable survival food as they are always available (although fall is usually the best time for collecting). The fresh rootstalk was usually peeled, then cut into sections and either dried, boiled or baked. If dried, the pieces were usually ground into a meal which was soaked in water, the water poured off — this being repeated several times to remove the bitterness.

Notes: Rootstalks were boiled, steeped in milk, and used to kill beetles and cockroaches. Scandinavian folklore states that there are trolls called "nacks" sitting on the bottom of lakes and ponds fishing for people. The long rubbery stem of the plant is their line and the flower is their lure. Watch out! The genus name comes from Nymphe, the Greek goddess of Spring.

Caution: Any plants harvested from present day waterways will have pathogens. Thoroughly wash and/or cook all plant parts collected in or around water. Use only small amounts of the rootstalks; large amounts could be poisonous.

Missing Link

Biologists have determined that the water lily may be the "missing link" in the evolution of flowering plants. It has been a mystery as to how, 150 or more million years ago, modern-day angiosperms (flowering plants) diverged from their closest relatives, the gymnosperms, (seed-bearing plants without flowers, such as cones from pines). Biologists found that the endosperm (the unique tissue which nourishes the embryo of the seed) of water lily have 2 sets of chromosomes. The endosperm of angiosperms have three sets and gymnosperms have 1 set. This could mean that water lilies could be an intermediate form linking these two.

Onagraceae Family | Fireweed | *Chamerion angustifolium* subsp. *circumvagum*

Fireweed habit

Fireweed fruit

Fireweed leaf with unique veination

Fireweed flower (lower left)

Evening Primrose Family

EVENING PRIMROSES

Oenothera spp.

Description: **Yellow evening primrose**, also called Hooker primrose (*O. elata* ssp. *hirsutissima*) is a tall, skinny biennial, mostly single-stemmed, from a strong taproot. Leaves are basal rosettes with stalkless leaves up the stem. Elongated clusters of flowers are bright yellow, 4-petaled, with 4 reflexed sepals, and a long calyx tube (hypanthium), blooming June to August. Found in dry, open areas in foothills and montane.

Stemless white evening primrose, also called tufted evening primrose, (*O. cespitosa*) is a low rosette of lance-shaped leaves. Large sweet smelling, snow white flowers up to 6" across emerge in the evening, then fade to pink and die the next day. 4 heart-shaped petals; 4 sepals reflex backwards; long calyx tube (hypanthium) looks like a flower stalk. Blooms from May to August. Found on dry, open sites, clay soils and roadsides from foothills to subalpine.

Medicinal uses: The oil of the seed of **yellow evening primrose** has found its place today as a natural supplement. The seed oil is high in the essential oil GLA (gamma linoleic acid) and is helpful in treating many problems like asthma, psoriasis, arthritis, weak immune system, premenstrual syndrome, heart and vascular diseases and for skin and nail health. The leaves and stems have an astringent and mucilaginous quality that can be useful as a tea, tincture or syrup with antispasmodic, nervine, sedative and muscle relaxant effects for coughs, sore throats, GI related depression, menstrual pain, hypersensitive nerves and as a healing stomach tonic. The plant infused in oil is a great wound healer and muscle relaxing remedy. The leaves and crushed seeds or roots of **stemless white evening primrose** were used by Native peoples as a poultice to reduce swellings, especially hemorrhoids, and a lotion for sores. All species share similar medicinal properties.

Edible uses: Because no evening primrose is known to be poisonous, try the different species in your area for their edibility. The roots of first year plants collected in the spring are the best, but in older plants they have an inner core that is edible. Harrington suggests many water changes throughout cooking as they can be very bitter. Young seed pods can be boiled, having a sassy, peppery taste. The seeds are roasted and ground into flour. The fresh flowers can be added to salads for a spicy flavor.

Notes: The **stemless white evening primrose** is pollinated by night flying insects (mostly moths) which have very long mouthparts to reach the nectar at the base of the long hypanthium.

Garden notes: Our **stemless white evening primrose** is hard to find in the industry, and even tougher to grow. It needs the worst soils, perfect drainage and very dry conditions. If you have a steep slope of shale, give this a try.

Special Pollinators

Notice the unique stigma on the stemless white evening primrose (top left picture), which has four branches forming an X shape. Pollination is by moths (particularly hawkmoths) and some bees, like most other flowers, but these pollen grains are held loosely together by viscin threads (a highly sticky substance). Only bees that are morphologically specialized to gather this pollen can effectively pollinate the flowers, plus the fact that these blooms open at night make these bees quite the specialists.

Louseworts

Orobanchaceae Family — *Pedicularis* spp.

Elephant head habit (left) and flower (above)

Giant lousewort flower

Fernleaf lousewort habit (above) and leaf (center inset)

Broomrape Family

Louseworts

Pedicularis spp.

Elephant Head

P. groenlandica

Other names: Little red elephant, elephantella, elephant's head lousewort

Fernleaf Lousewort

P. bracteosa

Other names: Bracted lousewort, wood betony, fernleaf

Giant Lousewort

P. grayi (*P. procera*)

Parrot's Beak

P. racemosa

Other names: Mountain figwort, sickletop lousewort, ram's horn

Description: All *Pedicularis* described below have deeply lobed, divided or toothed leaves, their flowers are bilabiate, the upper lip usually elongated into a narrow beak that arches downwards, sideways or upwards and all are found in moist areas.

Elephant head is 6-24" tall; often clustered stems with basal and alternate fern-like leaves. Pink-purplish to reddish (occasionally white) flowers shaped like an elephant's head with the flower beak slender and S-curved (the elephant's trunk) in a dense, spike-like cluster. Blooms June to August. Found in often dense colonies in wet meadows from montane to alpine.

Fernleaf lousewort is an erect leafy, unbranched perennial up to 2 feet tall. Leaves alternate, fern-like. Flowers pale yellow, not streaked with red, essentially beakless in a dense spike-like cluster. Blooms June to August. Found in moist to dry wooded sites from montane to subalpine.

Giant lousewort is an erect, leafy, unbranched perennial up to 3 feet tall. Leaves alternate, fern-like. Flowers pale yellow streaked with red, essentially beakless in a dense spike-like cluster. Blooms June to August. Found in shady, moist montane to subalpine forests.

Parrot's beak is an unbranched clump up to 20" tall with simple leaves, narrow, evenly and finely toothed. Leaves and stems often red-tinged. White flowers 2-lipped with a broad 3-lobed lower lip and an arching sickle-shaped lip like a down-curved beak. Blooms June to August. Found in subalpine conifer forests. (*See photos next page.*)

Medicinal uses: There is a long history of use of louseworts for a variety of internal problems like stomach ulcers, rheumatism and urinary problems. Michael Moore introduced this plant to me and I have come to really appreciate it as a muscle relaxant for muscles sore from over exertion. He called it the 'tight shoulder' remedy. It is also a nice sedative for kids and tranquillizer for adults. It stimulates sweating so can help reduce fevers. Its mild astringent and antiseptic properties make it useful for minor injuries. Michael used the whole flowering tops, but I find just the leaves of the plant in flower, fresh or dried, tea or tincture, work fine. I prefer the **fernleaf lousewort** and **giant lousewort** or a mixture of the two species. I occasionally mix in **elephant head** or **parrot's beak**. All of our species have similar properties but taste slightly different.

(*Continued next page.*)

Orobanchaceae Family — **Louseworts** — *Pedicularis* spp.

Parrot's beak habit
Note dark red stems

Parrot's beak leaf and flower

Broomrape Family

LOUSEWORTS
(CONTINUED)

Pedicularis spp.

Edible uses: The fleshy root can be eaten raw or cooked like carrots. The flowering stems, when young, are good as a pot herb. The flower nectar is a sweet treat and the flowers can be added to salads.

Notes: The *Pedicularis* are hemi-parasitic or semi-parasitic. They still have the benefits of green leaves for food manufacturing, but they depend on other plants for extra nutrition. They grow their tissues into unsuspecting hosts and share their energy. Many species have certain hosts they prefer: lupine, currants, groundsels (which are poisonous!). The toxins they obtain are then integrated into their own defenses. Historically, farmers in Europe believed that cattle feeding among these plants would become infested with lice.

Caution: Because these plants are semi-parasitic, they can pick up the characteristics of whatever plant they are attached to, including toxic substances. It's preferable to harvest under aspens and check surrounding plants for those with benign medicinal actions compatible to the louseworts, like oak, rose or raspberry. Because of the sedative nature and the trace presence of potentially toxic substances, moderation and experience must accompany their use.

The Plant Sucking Plants!

This would make a great horror flick: The Horrifying Plant Sucking Plant! Starring the louseworts and paintbrushes. These are hemiparasites, partially parasitic plants gaining strength from a host plant by attaching their roots to a host plant, while maintaining green leaves to make their own energy. In reality this is simply an insurance policy for the plant. During periods of drought the hemiparasitic plants have an advantage of extra water and nutrients, and they can expand their ecological niche into areas of poor quality soils. Paintbrush is probably the most well-known hemiparasite. Many paintbrush have almost no fine root hairs with which to absorb water and nutrients from the soil, showing their dependence on surrounding plants. Hiking through the meadows of the subalpine, it is pretty obvious that they take a particular liking to lupine and mountain currants. There always seems to be a riot of paintbrush color around these plants. At lower elevations, paintbrush are often found with sagebrush. The landscape trade is learning to grow paintbrush with a host to ensure better survival in our gardens. Louseworts have varied tastes for their host plants, and when harvesting you need to avoid areas with toxic plants such as senecios, lupines and goldenbanner because the lousewort accumulates the host toxins. Louseworts tend to be found in aspen and conifer sites.

Plantaginaceae Family — Plantains — *Plantago major* & *P. lanceolata*

Common Plantain fruit (left) and habit (above)

English Plantain habit (left) and narrow basal rosette (above)

Caution: Though rare, the plants may cause contact dermatitis. Avoid in patients with intestinal obstruction or abdominal discomfort.

Plantain Family

PLANTAINS
COMMON PLANTAIN

Plantago major

Other names: White man's foot, broadleaf plantain, rippleseed plantain

ENGLISH PLANTAIN

P. lanceolata

Other names: Ribwort plantain, buckhorn plantain, long-leaf plantain

Description: Both plantains are low perennial weeds growing from a basal rosette of leaves with distinctly parallel-looking veination. 4-petaled greenish-white flowers in a tight stalk on a long, narrow stem are followed by small green capsules, turning brown. Common in disturbed sites, lawns and waste sites from foothills to montane. **Common plantain** has broad, irregularly rounded leaves that tend to lay flatter to the ground, and the flower stalk is much longer and slender. **English plantain** has long, narrow leaves that are more erect, and wiry flower stalks almost a foot tall with stubby flower stalks and distinct white exserted stamens.

Medicinal uses: Plantains are safe and effective medicinal herbs. The leaves are used externally as a healing poultice and treatment for bleeding, quickly staunching blood flow and encouraging the repair of damaged tissue. They are one of my favorite outdoor first aid plants. The fresh plant chewed or crushed and placed on insect bites, stings and wounds is also mildly pain relieving, antibacterial and anti-inflammatory. More than once I have appreciated its drawing power to pull out splinters. The fresh leaves and leaf juice have been used externally worldwide for treating snake and insect bites, burns, bruises, rashes, pain, rheumatism, swellings, strained muscles, sprains, sore feet, and wounds. Internally, a decoction can be used to treat a wide variety of complaints including diarrhea, hemorrhoids, gastritis, peptic ulcers, vaginitis and upper respiratory tract inflammations. Heating the plant destroys the antibacterial effect. Plantain leaf ointment is great to stop all types of itching due to diaper and heat rash, poison ivy rash, eczema and dry skin. The seeds have been shown to reduce cholesterol. The seeds are high in mucilage and used as a source of natural fiber with lubricating and bulking laxative effects. *Plantago psyllium*, grown in South America, is the source of psyllium seed husks found in Metamucil. These plants grow everywhere, probably right in your lawn or driveway. Learn to use them and love them…they can be your healing friends.

Edible uses: You have to get this one really young to enjoy it. The leaves can be used like spinach. By late spring the leaves are so tough it is pretty chewy to eat. Some people will put them in soups and stews to get the nutritional advantage. The ripe seeds can be ground into meal or flour for bread or pancakes.

Notes: The rough strings in the leaf have been used as a fiber source. The leaves used as a hair rinse to treat dandruff. Native Americans called plantain the "white man's boot" or "footprint" because they noticed how prolifically it grew around the new European settlements. The generic name comes from the Latin planta meaning "sole." Plantain makes a good breath freshener, probably because of its high chlorophyll content.

Garden notes: Although most gardeners mercilessly root them out of their lawns, they actually do no harm when growing there and, indeed, help to maintain the fertility of the lawn. Their presence can indicate heavy clay soils that are waterlogged or poorly drained. They are an important food plant for the caterpillars of many species of butterflies.

Polygonaceae Family · Sheep sorrel · *Rumex acetosella*

Sheep sorrel arrow-shaped leaf

Sheep sorrel flowering (above) and flower close-up (top right)

Sheep sorrel flowering habit

SHEEP SORREL
Rumex acetosella (Acetosella vulgaris)

Other names: Sheep's sorrel, red sorrel, field sorrel, schav

Description: Slender rhizomatous perennial European weed up to 12" tall with arrow-shaped leaves. Spring growth is a messy rosette of these uniquely shaped leaves. Flowers reddish (sometimes yellowish), either male or female on long, narrow spikes followed by small, shiny yellow-brown fruits enclosed in papery calyxes. A clump of flowers from a distance is distinctly reddish. Found on disturbed and waste sites and in gardens from plains to subalpine.

Medicinal uses: This plant was commonly used in Europe as a cooling diuretic and poultice for sebaceous cysts, skin cancers and tumors. Leaf teas were used for fever, scurvy and inflammations. Roots were used for diarrhea and excessive menstruation. The fresh juice of the leaves is diaphoretic, diuretic and mildly laxative. The plant is also part of a North American formula called Essiac which is a popular treatment for cancer. Its effectiveness has never been reliably proven or disproven since controlled studies have not been carried out. The leaf juice is useful in the treatment of urinary and kidney diseases. Sheep sorrel is considered a detoxifying herb acting on the blood and liver with antibacterial and antifungal properties.

Edible uses: The spring rosette is a tasty lemony, vinegary treat! The leaves, up until the flower stalk forms, are an excellent thirst-quenching trail munchie, and can be used as a potherb. The plant is high in Vitamin C, beta carotene, potassium and phosphorous. Because of the high amounts of oxalic acid, it is a good idea to combine it with dairy products if you are consuming a lot of it. A sorrel infusion is nice chilled with sweetener for a cooling drink.

Notes: Roots produce dark green to brown dyes. The word "sorrel" comes from the French *sur*, meaning sour. The scientific name *acetosella* means "little vinegar plant." The common name, sheep sorrel, refers to the leaf shaped like a sheep's head with ears.

Garden notes: One of the host plants for the American Copper butterfly. The presence of sheep sorrel in our gardens often indicates sandy, light soils with acidic or low lime conditions. Farmers say that where they see patches of sheep sorrel spreading, the soil is worn out.

Caution: Plants can contain quite high levels of oxalic acid, which gives the leaves an acid-lemon flavor. The leaves should not be eaten in large amounts since the oxalic acid can lock-up other nutrients in the food, especially calcium, thus causing mineral deficiencies. Cooking reduces the oxalic acid content. People with a tendency to rheumatism, arthritis, gout, kidney stones or hyperacidity should take special caution if regularly including this plant in their diet since it can aggravate their condition. The plant may be poisonous to livestock due to the oxalate acid content.

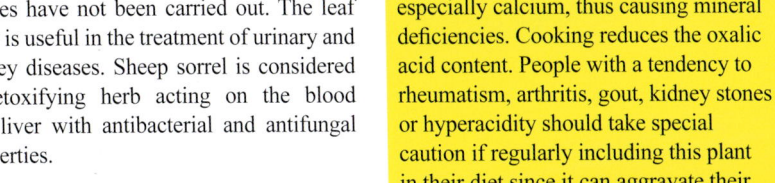

Sorrel Sauce for Seafood or Cooked Vegetables

1 pint wood sorrel or sheep sorrel leaves   *2 tbsp butter or olive oil*

Melt the butter over low heat in a medium pot. Add the sorrel and stir until the sorrel leaves are wilted. Spoon the sauce over the seafood or cooked veggies. Sorrel sauce is also a great soup base. You can freeze sorrel sauce. *Serves 2; recipe can be multiplied.*

— *Mother Earth News*

Polygonaceae Family Bistort *Bistorta bistortoides*

Bistort habit

Bistort leaf and flower

Buckwheat Family

Bistort

Bistorta bistortoides
(*Polygonum bistortoides*)

Other names: American bistort, snakeweed, bottlebrush, knotweed, smartweed, western bistort

Description: A very slender perennial growing 8-27" tall. Stem leaves are narrow, smaller than the mostly basal leaves, and sessile. Basal leaves are stalked and up to 10" long. Flowers form a fluffy, tight white or pink-tinged cluster atop a single stem. Stamens exserted. Found in moist meadows and near streams from montane into the alpine. Blooms June to August.

Medicinal uses: The strong astringent and disinfecting roots are used externally dried and powdered and applied to cuts and wounds to stop bleeding and prevent infection. Taken internally it is excellent for bleeding, such as from nosebleeds, heavy periods and wounds. The root has been used for treating many conditions from diarrhea and dysentery to cholera, colitis, irritable bowel syndrome and peptic ulcers. It can be used as a wash, tea, douche, salve and powder. This is a good first aid ally to know.

Edible uses: This is a really weird looking root (think of the name snakeweed) that is actually pretty tasty. If you find one low in tannins, they are downright pleasant, somewhat like water chestnuts, kind of nutty. If you get a bitter one, I recommend moving on to another; these can contain up to 20% tannins which will not be friendly to your stomach. Take a scrubber and try and rub the rough outer layer off if objectionable. I find it doesn't make any difference to the taste. These roots can be eaten raw, preferably the younger ones. Older roots are bitter and fibrous. Roots can be boiled, roasted, baked, dried and ground into flour. Leaves and shoots have a nice tangy flavor cooked as a potherb or eaten raw (stay with younger parts). The brown seeds can be roasted and added to thicken stews, or ground into flour.

Notes: You might recognize the buckwheat family from its more common edibles; buckwheat and rhubarb.

Bistort Pudding

Bistort leaves
Dandelion leaves
Nettle leaves
1 oz boiled barley
1 hard-boiled egg
½ oz butter
Pepper and salt

Pick only the young leaves of bistort, dandelion and nettle and cover them with boiling water for 20 minutes. Drain off the water and chop the leaves. Add the boiled barley to the mixture, together with the chopped hard-boiled egg and butter. Season with salt and pepper and stir well. Reheat and place in a pudding basin.

Nature's Wild Harvest, Eric Soothill & Michael J. Thomas, 1990

Polygonaceae Family — Buckwheats — *Eriogonum* spp.

Sulphur flower habit

Buckwheat fruit

Subalpine buckwheat flower (oval) and habit (above right)

Buckwheat Family

BUCKWHEATS SULPHUR FLOWER

Eriogonum umbellatum

Other names: Sulfur buckwheat, wild buckwheat, umbrella plant, sulfur plant

SUBALPINE BUCKWHEAT

Eriogonum umbellatum var. *majus* (*Eriogonum subalpinum*)

Other names: Wild buckwheat, umbrella plant, sulfur plant, false buckwheat, creamy buckwheat

Description: Both species are lovely mat-forming plants up to 2 feet wide. Leaves are all basal, leathery, green above and densely hairy below. Single stems up to 12" tall carry an umbel of bright yellow (Sulphur flower) or creamy turning to rose (Subalpine buckwheat) flowers subtended by leafy bracts. Both are found in dry, exposed sites from foothills into the alpine. Blooms June to July.

Medicinal uses: Our local buckwheats have astringent properties that are useful in treating diarrhea and for shrinking and soothing inflamed membranes and slowing secretions from irritated membranes. The plant tea can be used as a soothing eyewash, douche, gargle, enema or sitz bath. It seems to have a special affinity for soothing lower urinary tract infections. The tea is mildly hemostatic and can be used to decrease menstrual and postpartum bleeding. Externally, like all astringents, it is good as a wash for rashes, cuts and sores. The plants were used in steam baths for rheumatism, stiff joints and muscles, general illness and as a ceremonial purification remedy.

Edible uses: Stems were eaten both raw and cooked before flowering. I have used seeds from subalpine buckwheat as a nice addition to cereals and pancakes. After collecting, I separate the seed from the chaff and then grind them.

Notes: All buckwheats are valuable honey plants.

Garden notes: The hottest, driest places are where the buckwheats excel. I use them for erosion control on really inhospitable slopes. Their dense mats catch organic debris which decays. They will also spread by reseeding. If you are transplanting, choose the smallest plants you can find and take a large section of soil, getting the entire root if you can. Seeds also work, but are rather slow.

Caution: Not recommended during pregnancy due to the plant's potential effect of vasoconstriction of the uterine lining.

Cryptobiotic Soils

Our sagebrush-steppe communities have a greater diversity of plant life than in most adjacent habitats. In years with above average precipitation, a flurry of wildflowers color the gray sagebrush flats. Looking closely on the soil surface one finds a living soil, cryptobiotic crust, consisting of fungi, lichens, green algae, brown algae and possibly mosses growing in a highly intertwined living structure. This crust stabilizes soil, retains moisture and keeps non-natives from encroaching. Grazing and human impacts destroy this protective crust, which may never recover its structure or function. Please respect this valuable and fragile living system!

Polygonaceae Family **Yellow dock** *Rumex crispus*

Yellow dock leaves

Yellow dock fruit

Caution: Yellow dock root should not be taken in cases of bowel and gallbladder obstructions and in cases of known history of oxalate kidney stones. Plants can contain quite high levels of oxalic acid, which gives the leaves an acid-lemon flavor. The leaves should not be eaten in large amounts since the oxalic acid can lock-up other nutrients in the food, especially calcium, thus causing mineral deficiencies. Cooking reduces the oxalic acid content. People with a tendency to rheumatism, arthritis, gout, kidney stones or hyperacidity should take special caution if regularly including this plant in their diet since it can aggravate their condition.

Buckwheat Family

Yellow Dock
Rumex crispus

Other names: Curled dock, curly dock, sour dock

Description: A robust perennial weed from Europe of strap-shaped leaves up to 2 feet long in a rosette from a long, stout yellow taproot (hence the name yellow dock). Flower stalk up to 5 feet tall with smaller alternate leaves and clusters of tiny green, white or pinkish flowers from the leaf axils. Seeds are rusty-brown papery winged clusters. There could be up to 40,000 seeds per plant! Found in moist, disturbed areas from plains to montane. Flowers June to August.

Medicinal uses: Yellow dock has been used world-wide as medicine for centuries and was listed in the U.S. Pharmacopeia from 1863 to 1905 and the National Formulary from 1916 to 1936. Other docks may have similar properties but yellow dock is the most widely used. The whole plant can be used but the root is the most medically active and the darker yellow the root, the stronger the medicine. It is used internally to treat constipation, diarrhea, bleeding of the lungs, blood disorders, chronic skin diseases, rheumatism, jaundice, hemorrhoids and indigestion. Externally, the leaves or mashed root as a poultice, salve, wash or dried as a dusting powder for sores, wounds, ulcers, swellings, itching, and other skin problems. Today, yellow dock is used as a liver stimulant, especially good for poor digestion of fatty foods, and a blood cleanser. It is being used to treat jaundice and post-hepatitis flare-ups, swollen lymphatic nodes, sore throats, skin sores, warts and rheumatism. The root also has antifungal and antibacterial actions. It is an essential ingredient in spring cleansing formulas along with dandelion and burdock roots. There is some evidence that root preparations will help the body in eliminating concentrations of lead, arsenic, and other heavy metals.

Edible uses: The early spring leaves are a tangy treat cooked as a potherb. By mid-spring they will set your teeth on edge with their bitterness. The young leaves can be eaten raw; best mixed with more neutral greens. Do not eat the young leaves raw without washing them first as they have chrysophanic acid which will irritate your mouth for hours. The flower stalk in mid-spring has a tangy lemony flavor. Cut them off at the base, peel off the outer layer, add them raw to salads or cook them for about 5 minutes. I find the seeds too labor-intensive to harvest, although they have been used historically by separating the outer hull from the small seed and boiling them or grinding them into flour. Some tribes leached the seeds before using them, much like acorns were leached.

Notes: Roots produce dark green to brown and dark grey dyes. *Dock* is the term for the solid part of an animal's tail, and *to dock* means to remove a tail. All "weeds" were eventually called "dock" in English, maybe because they were trying to remove, or *dock*, them.

Garden notes: It is considered to be a serious weed of agriculture, but it is a good plant for re-establishing fertility in the soil. Its deep roots bring up nutrients that would otherwise be lost while its leaves make excellent compost. It is also a very important food plant for the caterpillars of many species of butterfly. It is an ingredient in herbal compost activator formulas. If growing in our gardens, they indicate waterlogged or poorly drained soils with acidic or low lime conditions.

Pipsissewa

Pyrolaceae Family — *Chimaphila umbellata* spp. *occidentalis*

Pipsissewa mature flower and fruit

Pipsissewa flowering

Pipsissewa habit

Wintergreen Family

Pipsissewa

Chimaphila umbellata ssp. *occidentalis*

Other names: Prince's pine

Description: Semi-woody evergreen sub-shrub up to 12 inches tall with whorls of 3-8 thick, dark green, kite-shaped, toothed leaves. Flowers light pink or rose tinged, nodding in an umbel above the leaves, producing fruits of dry, round capsules. Blooms June to August. Found in (usually) coniferous forests from foothills to montane.

Medicinal uses: Pipsissewa was an important herb among Native Americans, who used it for various problems, including head colds, backaches and rheumatism, gonorrhea, lung congestion and mixed with other herbs as a blood purifier. The early settlers used it as a tonic and diuretic for kidney complaints and rheumatism. It is mainly used in an infusion for urinary tract problems such as cystitis and urethritis and to induce sweating to lower fevers. It has also been prescribed for more serious conditions such as gonorrhea and kidney stones. By increasing urine flow, it stimulates the removal of waste products from the body and is therefore helpful in treating rheumatism and gout. It is also a lymphatic catalyst. The fresh leaves may be applied externally to rheumatic joints or muscles, as well as to blisters, sores and swellings. A homeopathic remedy is made from the leaves. It is used in the treatment of inflammations of the urinary system.

Edible uses: Its pleasant taste has been used to flavor candies, soft drinks (especially root beer) and beer. The leaves and berries have been used as trail munchies, but I find them very bitter. It has been overharvested in areas to the point of extermination. This is a stunning plant that we should enjoy and not collect.

Notes: The plant is used in perfumery because of its delicate scent.

Garden notes: Although not common in the landscape industry, pipsissewa makes a beautiful groundcover for coniferous shade areas in our landscapes. I have never tried transplanting them as I hate to take the plants, the few that there are in the wild, from their homes in nature.

> **Caution:** The fresh leaves, moistened, are a counterirritant and if applied to the skin for long periods of time, half-hour or longer, can cause skin irritation, blistering and peeling. In tests on animals, pipsissewa leaves appear to lower blood sugar levels. It is not suitable for long term internal use as it may cause diarrhea, nausea and vomiting and reduce mineral absorption from the gut.

The Soil is Alive!

One very important group of living organisms is a group of fungi called mycorrhizae. These fungi form a symbiotic relationship with myriad plant roots, assisting in converting otherwise insoluble nutrients from the soil and even rocks, into forms available to the plant. In return, the fungi are afforded carbohydrates from their host. These relationships are especially important in low-nutrient soils.

Rosaceae Family — **Potentilla** — *Potentilla* spp.

Potentilla gracilis flower (left) and leaf (below)

Potentilla pulcherrima habit (lower left); note the silver undersides of the leaves

Drymocallis (Potentilla) arguta with pinnate leaves and cream flowers

Rose Family

POTENTILLA

Potentilla spp.

Other Names: Cinquefoil, five-finger

Description: There is a dizzying variety of native, and a few non-native, potentilla that can give a botanist a headache trying to key them out. The common name cinquefoil means 5-leaf, referring to the typical 5-7 palmate or pinnate leaflets (although there can be up to 30 or more). Many have silver hairs on one or both sides of the leaf. Each flower has sepal-like bracts beneath its sepals and yellow petals (with a couple of exceptions). All are perennial herbs except for one that is a shrub. This shrub is called shrubby cinquefoil (*Pentaphylloides floribunda*, also known as *Potentilla floribunda*). They also can be found in virtually every habitat with many of them being excellent pioneer species growing in disturbed areas.

Medicinal uses: *Potentilla* means "little potent one" referring to their effectiveness in stopping bleeding and dysentery. Like all members of the Rose Family these have astringent constituents that help to tighten and tone tissues and decrease inflammation. The herb or root can be made into a tea or tincture and used as a gargle or mouth wash for sore throats, tonsil and gum inflammation, to decrease diarrhea, internal bleeding and fevers, speed healing of esophageal and stomach ulcers. Topically, the leaf tea can be used as a wash for sunburns, wounds, poison ivy and oak and other skin irritations. Placing leaves of potentilla in your boots on a long hike can help prevent blisters. Because of its abundance, this is a good plant to know for trail first aid. All species work similarly.

Edible uses: Although the potentillas are in the Rose Family of highly edible plants, I have only tried one. Silverweed, *Potentilla anserine* ssp. *anserine* (*Argentina anserina*) is a fairly common plant with tasty roots either raw or boiled. They are rather small, but have a nice nutty flavor. Silverweeds are easy to identify by their solitary yellow flowers, distinct grayish leaves, and runners.

Notes: A tea was used to prevent saddle sores on horses.

Garden notes: There are many highly valued landscape plants in this genus. Shrubby cinquefoil (shrubby potentilla) is a long-blooming versatile shrub that can take a little shade, but can also take full sun and dry conditions.

> **Caution:** Make sure you have correct identification as the flowers of potentilla can look like buttercups, most of which are poisonous.

Rose or Buttercup?

Considering most of the Rose Family members are highly edible and the Buttercup Family members are considered toxic, it is good to understand some solid ID traits that separate the families. Rose Family members have a very distinct hypanthium, a cup-shaped structure at the base of the flower. It is more obvious in larger flowers, but can still be seen as a fleshy "disc" at the base of smaller flowers. The petals, sepals and stamens arise from the rim of the hypanthium and the hypanthium is lined with nectaries which produce nectar. Buttercup Family members do not have this large fleshy base or disc. Rose Family members also have a stipule, a small pair of leaf-like structures at the base of a leafstalk. Buttercup Family members lack this stipule.

Rosaceae Family — Strawberry
Fragaria vesca ssp. *bracteata*
Fragaria virginiana ssp. *glauca*

Wild strawberry flower, leaf (above) and red fruit (oval left)

Caution: Partially wilted leaves can contain toxins. Always use fresh or completely dried leaves. Some people develop a rash or hives after ingestion of large quantities of the fruit.

Woodland strawberry

Rose Family

STRAWBERRY

WOODLAND STRAWBERRY

Fragaria vesca ssp. bracteata

WILD STRAWBERRY

Fragaria virginiana ssp. glauca

Description: Our two species are ground hugging perennials of small clumps with characteristic red stolons creating new plants. Leaves are trifoliate with sharply toothed leaflets. The white flowers are 5-petaled with many stamens, forming loose clusters. Fruits are tiny little seeds embedded in a red, fleshy receptacle (the strawberry). Woodland strawberry is typically found in moister sites than wild strawberry, both up to subalpine. Blooms May to August.

Medicinal uses: With the gentle astringency and high mineral content of all the Rose Family, the leaf tea of wild strawberry can be used as a tonic for the women's reproductive system with or in place of raspberry leaf. The leaf tea can be used to soothe inflammations of the mouth, eyes and skin. This tea has a long history of use for treating diarrhea, fevers, dysentery, and kidney problems and as a blood tonic, stomach cleanser and valuable source of vitamin C. The root is astringent and diuretic for treating diarrhea and chronic dysentery and an aid in treating liver and kidney complaints. The fruit is considered cooling and diuretic and used in Europe to treat tuberculosis, gout, arthritis and rheumatism. Mashed berries on the skin are cooling and soothing to sunburn.

Edible uses: The tiny berries may be small, but are bursting with flavor. The trick is to find them hidden under the leaves, and get to them before all the other animals. A handful of fresh, bruised leaves or berries steeped in hot water make a flavorful tea high in iron and Vitamin C. A traditional method of preparing them involved mashing the berries, spreading them over grass or woven mats and drying into cakes. The cakes can be eaten dry or re-hydrated. The flowers, leaves and stems were mixed with various roots in cooking pits to add flavor.

Notes: Hold the juice of strawberries in your mouth for a few minutes, then rinse with warm water to remove tartar and whiten teeth.

Garden notes: Wild strawberries will give you lots of bang for your buck. Wild strawberry provides erosion control on drier sites with their trailing habit, (they don't like totally nasty soils). Woodland strawberry is used as a ground cover under shade trees.

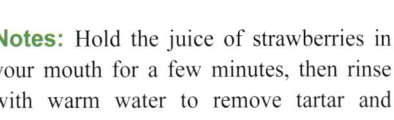

"Are wild strawberries really wild? Will they scratch an adult, will they snap at a child? Should you pet them, or let them run free where they roam? Could they ever relax in a steam-heated home? Can they be trained to not growl at the guests? Will a litterbox work or would they make a mess? Can we make them a Cowberry, herding the cows, or maybe a Muleberry pulling the plows, or maybe a Huntberry chasing the grouse, or maybe a Watchberry guarding the house, and though they may curl up at your feet oh so sweetly, can you ever feel that you trust them completely? Or should we make a pet out of something less scary, like the Domestic Prune or the Imported Cherry? Anyhow, you've been warned and I will not be blamed if your Wild Strawberries cannot be tamed."

— Shel Silverstein, *Where the Sidewalk Ends*

Rubiaceae Family Northern bedstraw *Galium boreale*

Northern bedstraw flowering habit

Northern bedstraw's smooth whorled leaves around square stem

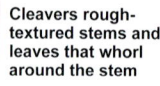
Cleavers rough-textured stems and leaves that whorl around the stem

Madder Family

Northern Bedstraw

Galium boreale
(*G. septentrionale*)

Other names: Bedstraw

Related species: Cleavers (*Galium aparine*) has rough textured stems and leaves (they feel like sandpaper when rubbed). Do not eat these raw or they will stick in your throat (Yuck!). The roots of cleavers produce a purple dye.

Description: A weak stemmed perennial with whorls of 4 leaves along the stem, and topped by fragrant 4-petaled white sprays of flowers. Fruits are hairy nutlets in pairs. Found in drier forests from montane to subalpine. Blooms June to August.

Medicinal uses: All species of *Galium* share similar medicinal properties. The whole plant tea is diuretic, diaphoretic and contraceptive. It has been used for various problems such as lymphatic swellings, fluid retention, fevers, kidney inflammation and stones, cystitis and mouth ulcers. Topically it is good for skin problems like poison ivy, rashes, eczema, psoriasis, burns and inflammations. In folk medicine it was used as a blood purifier, cancer remedy and diuretic.

Edible uses: The youngest shoots can be cooked like spinach, and the young leaves are nice in salads. Also boil the young shoots, cool and place in salads. The dried leaves make a pleasant tea. The small seeds can be roasted to a dark brown then ground to make a coffee substitute (they are in the Coffee Family after all).

Notes: Bedstraw earned its name from the light-weight plant forming a sweet smelling, insect repelling mattress stuffing. The name Galium comes from the Greek gala, milk, which is curdled by some species of bedstraw.

Garden notes: A non-native species, sweet woodruff (*Galium odoratum*) is a smaller, though tenacious, ground cover with charming aromatic, star-like flowers in abundance. It will quickly cover shade areas that are difficult to get anything else to grow.

Sleeping in the Forest

I thought the earth remembered me, she
took me back so tenderly, arranging
her dark skirts, her pockets
full of lichens and seeds. I slept
as never before, a stone
on the riverbed, nothing
between me and the white fire of the stars
but my thoughts, and they floated
light as moths among the branches
of the perfect trees. All night
I heard the small kingdoms breathing
around me, the insects, and the birds
who do their work in the darkness. All night
I rose and fell, as if in water, grappling
with a luminous doom. By morning
I had vanished at least a dozen times
into something better.

— Mary Oliver

Scrophulariaceae Family — Figwort — *Scrophularia lanceolata*

Figwort leaf and square stem

Figwort habit (below right)

Figwort flower

Figwort Family

FIGWORT

Scrophularia lanceolata

Other names: Bunny in the grass, lanceleaf figwort, hare figwort, American figwort

Description: Tall erect plant, growing 2-7 feet tall and 2-4 feet wide. Spear-shaped, irregularly or double serrated leaves are arranged opposite the square stem; leaves can reach up to 8" long. Small irregular flowers up to ½", the petals reduced to a single, small bilobed upper petal (like bunny ears!); colored from red, yellow and orange, to brown and green; arranged in a highly branched panicle. Found in often rocky and/or disturbed moist sites from montane to subalpine. Blooms from late June through summer.

Medicinal uses: The above ground plant, fresh or dried in teas, tinctures, washes, salves and poultices is antifungal, astringent, sedative and diuretic. It has been used to treat everything from athlete's foot, cradle cap, eczema, rashes, burns, bruises, recent joint injuries, hemorrhoids, hives, arthritis and the list goes on. It supports detoxification and can be helpful for folks with chronic low grade skin problems. Check out Michael Moore for a lengthy discussion of its uses. This is a plant worth getting to know and cultivating since it is now considered "at risk" in New Mexico and its range seems to be decreasing in Colorado.

Edible uses: None known

Notes: This species is primarily pollinated by wasps that have been found to be 7 times more efficient at pollinating this genus than bees, who also visit the plant for its nectar. Most wasps do not eat pollen but do feed on nectar; many wasps are predatory and feed insects to their larvae. The Xerces Society recommends this plant for its value to native bees and wasps. The genus name *Scrophularia* was given by Linnaeus in 1753. It is Latin for "a tumor or glandular swelling" and was thought to be a cure for them.

Garden notes: Because I see them so rarely in our area I have not tried to transplant them into my garden, although I think they would take from seed quite well. There is a stunning *S. macrantha* available in nurseries now with bright red "bunny" flowers that has thrived in my garden. We can find no known medicinal use for this species, but feel it is a worthwhile addition to a native garden.

Caution: Not to be taken internally during pregnancy or by those with heart conditions.

Unpopular Pollinators

Most sources defining the pollinators for the Scrophularia talked about social wasps. Interesting! Wasps really are not popular pollinators as pollen doesn't stick to hairless and hard-shelled insects (wasps are NOT fuzzy and cute like most bees!). Wasps also feed their larva by preying on other insects rather than collecting nectar and pollen; they are pretty lazy pollinators. But figworts seem to be well suited to wasp pollinators. The dreary color and foul smell could be a draw for wasps looking for insect nutrition and whose predatory instinct leads them to the flower. From the plant's point of view this specialization has advantages: the plant does not need to compete with other plants for pollinators, insects don't tend to visit these flowers so pollination is more assured as the wasp travels from flower to flower.

Scrophulariaceae Family — Mullein — *Verbascum thapsus*

Mullein first year leaf rosette

Mullein habit (far left) Flowers and budded stalk (center oval)

Mullein with unusual crestate stalk

Mullein second year leaf rosette starting to form flowering stalk (left)

Figwort Family

MULLEIN
Verbascum thapsus

Other names: Great mullein, common mullein, velvet dock, flannel mullein

Description: A Eurasian biennial weed very conspicuous during both years of growth. The first year is the characteristic fuzzy rosette of large (up to 12") broadly lance-shaped leaves. The second year sprouts a tall spike, many exceeding 6 feet, of densely packed yellow flowers (looking like a corn cob). Found in disturbed areas from foothills to subalpine. Blooms June to August.

Medicinal uses: Mullein has a long history of use and is still a popular medicinal today. It is gentle, safe and treats a diversity of conditions. The leaves are used for treating a variety of respiratory complaints like dry coughs, congestion, asthma, TB and bronchitis by loosening and expelling mucous. The leaf infusion should be strained to remove the fine hairs. The leaves can be used as a smoke inhalant or in smoking mixtures for asthma, to calm respiratory spasms and for fevers. Kidney problems, diarrhea, dysentery and lymphatic stagnation are also treated with leaf teas. Externally, leaf poultices are used for problems like ulcers, tumors, hemorrhoids, swollen glands, joint pain and arthritis. The root in tea or tincture can stimulate urination while toning the urinary system in cases of incontinence, bed-wetting and early prostatitis. The infused flower oil is probably the best known mullein remedy for treating ear infections. It is anti-bacterial and decreases pain while softening the wax and allowing for movement of built up fluids. Almost all body systems can benefit from this plant and it should be a part of every herbalist's medicine cabinet. A homeopathic remedy is made from the fresh leaf and flower.

Edible uses: I don't really consider this plant "palatable" because of the dense hairs, although the leaves can be used as tea: be sure to strain thoroughly.

Notes: The uses of mullein go back to the Romans who would dip dried flower stalks in tallow to use as torches. The plant has been used for yellow dye, hair dye, insecticide, insulation and tinder.

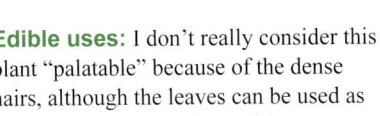

Caution: The seeds should not be consumed. Although no side effects have been noted, the leaves contain tannin as well as small amounts of coumarin and rotenone — two substances that could be toxic if consumed in larger quantities. The leaf fuzz may irritate sensitive skin and throat membranes.

Prolific and Long Lived Traveler

Mullein is native to Europe, northern Africa, Asia and western China. It was brought to the United States in the 1700's for its medicinal properties and use as a fish poison. Today, it is an established weed in North America, Australia, New Zealand, tropical Asia, Hawaii, Japan and several South American countries. Each plant produces hundreds of fruit capsules, each containing up to 700+ minute seeds. The seeds can remain viable for decades up to a hundred years, according to some studies. They remain in the soil seed bank for extended periods of time, and can sprout from apparently bare ground, or shortly after forest fires, long after previous plants have died. Mullein rarely establishes on new ground without human intervention because its seeds do not disperse very far. Seed dispersion requires the stem to be moved by wind or animal movement; 93% of the seeds fall within 15 feet of the parent plant.

Typhaceae Family — Cattails — *Typha latifolia*

Cattail leaves and brown female stalks going to seed

Left: Brown pollinating male atop green female stalks

Cattail habit

Cattail Family

BROAD-LEAVED CATTAIL

Typha latifolia

Other Names: Common cattail, reedmace, broadleaf cattail

Description: Emergent (a plant partially submerged in water) perennial with stout stems up to 5 feet tall from coarse rhizomes. Leaves alternate, grass-like, stiff and rather spongy. Flowers in dense spikes with yellow male flowers (with pollen) at the tip (disintegrating quickly) and immediately below are the green, turning brown, female flowers. Seeds attach to fluff of hairs; the whole spike looks like cotton when mature. Found in still or slow moving water from foothills to montane. Bloom June to July.

Medicinal uses: Rarely used as medicine today, it was used extensively by Native peoples. The rootstalks were used as a poultice for burns, wounds and inflammations. The young flower heads are astringent and were eaten to relieve diarrhea and other digestive disorders. The root can be infused in milk or water for a tea to treat dysentery and diarrhea and abdominal cramping. The down from the flowers can be used as a wound dressing. Pollen is used for flower essence. Chinese herbalists use the pollen as an astringent to stop bleeding, treat diarrhea and kidney stones.

Edible uses: Young shoots in early spring: Pull the vertical shoots from the rootstalks when they are 4-16" long. Peel outer leaves, leaving the inner white section. Eat raw, put in salads or use as a potherb (called Cossacks asparagus). You can use the shoots up to 2 feet tall then they become very fibrous. Boil these older shoots to make them more palatable.

Young flower spike: Take the young green flower spikes out of the sheath and boil for 20 minutes and eat like corn on the cob. Or scrape off the immature flowers and use as a flavoring and thickening.

Flowers further along, but not yet producing pollen: Strip off the pollen producing section (top part of spike) and mix with wheat flour to make muffins, pancakes, cookies, etc. These flowers can also be dried.

Pollen: As the top part of the spike produces pollen, put a paper bag over the spike and give the stem a good shake. The pollen will gather in the bag. Separate the pollen from the insects, etc. by shaking. Pollen is highly nutritional and used as a delicious addition to other flours.

After tops have browned: Now is the time to use the starchy rootstalks, although they can be used at any time; late summer will produce a starchier rootstalk. Dig into the mud for the rootstalk which can be dried for flours and used in soups and stews. At the end of the rootstalk is a young bud. Use your thumb to snap off the bud. Remove the outer peel, use the central white core. Some people say it can be eaten raw, but with all the nasty stuff in water any more I always cook them by boiling or baking.

Seeds: Roast or parch the spike then rub the seeds off. They are small but very nutritious.

Notes: The starchy white root core was boiled to make syrup which was fermented to make ethyl alcohol. Rootstalks and leaves were used as caulk for boats and barrels. Quilts made from cattail down are water repellent. Cattail down mixed with lime and ashes make cement as hard as marble. Pollen was historically used in religious ceremonies, to be replaced by corn pollen. Cattail seed is said to kill mice. The stems were used to make a type of glue. During WWI, the fluff, called Kapok, was stuffed into life jackets and sleeping bags

Typhaceae Family — Cattails — *Typha latifolia*

Cattail collecting

Cattail seed

Cattail Family

Broad-Leaved Cattail (continued)

because it floats and is waterproof. Leaves were used for mats, weaving, sandals, and baskets. Down was used for stuffing diapers and in baby beds, and as fire starters.

Garden notes: If you have a pond or wetland area, you will eventually have cattails. Their fluffy airborne seeds cover a lot of distance. To some they are a blessing, providing food, shelter and nesting sites for numerous birds and critters. To others they are an unwanted invasion. If you want to introduce them, dig the rootstalks making sure to have a couple of healthy shoots with buds. Take a coat hanger and pin it down into the mud or it will float to the surface. Research has proven their effectiveness for cleaning polluted and highly mineral waters (from mine tailings).

Caution: Because cattails grow in water of dubious origins, it is highly recommended that all parts be thoroughly cooked before eating. Collect from pollution free sites.

Cattail Shoots with Wild Mint

1 lb. trimmed cattail shoots, cut into 4 inch lengths
2 cloves garlic, chopped fine
6 Tbsp. extra-virgin olive oil
1/4 cup fresh wild mint leaves, chopped

Steam shoots until tender-crisp. In small skillet cook garlic in oil until pale golden. Toss shoots with garlic mixture, chopped mint, and salt and pepper to taste. Serve warm or at room temperature. Serves 4

— From *The Rocky Mountain Wild Foods Cookbook*

Cattail Corn Bread

1/4 cup sugar
1 Tbsp. baking powder
1/2 tsp. salt
1/2 cup cattail pollen
1/2 cup yellow cornmeal
1 cup all-purpose flour
1 egg
1 cup buttermilk
1/3 cup vegetable oil

Put dry ingredients in mixing bowl. Add remaining ingredients and mix until ingredients are just moistened. Spoon into oiled 8x8 inch pan. Bake in preheated 425°F oven for 15 to 20 minutes.

— From *The Rocky Mountain Wild Foods Cookbook*

Stinging nettle

Urticaceae Family — *Urtica dioica*

Stinging nettle leaf and flower

Stinging nettle flowering habit (left)

Stinging nettle young plant habit

Nettle Family

Stinging Nettle

Urtica dioica

Other names: Nettle, slender nettle

Description: A leafy stemmed erect perennial up to 2 feet tall. Stem is four sided. Leaves are opposite and simple with roughly serrated edges and stinging hairs. Flowers are green or pinkish clusters hanging from the upper leaf axils. Found in moist, rich, often disturbed sites from foothills to montane.

Medicinal uses: I really appreciate this plant! The aerial tops harvested before flowering are used as a nutritive tonic, high in iron, minerals, vitamins A and C, and a diuretic and urinary tract astringent. It is anti-allergenic and a natural anti-histamine used to treat hay fever, pollen and food allergies, asthma, itchy skin conditions and insect bites. It also reduces bleeding associated with wounds, menstruation, bladder infections and hemorrhoids and increases mother's milk flow. Other applications include kidney stones, premenstrual syndrome, benign prostatic hypertrophy, gout, multiple sclerosis, dental plague, diarrhea, sciatica, arthritis, many respiratory conditions, Alzheimer's and anemia. This is an amazing healing plant and should be in everyone's home remedy collection.

Edible uses: The stinging hairs are destroyed by thorough heating and drying. Collect the young leaves, especially in the spring, boil them and use like spinach. Avoid using the plant after flowers have appeared as hard particles called cystoliths are present and irritate the urinary tract. Use the nettle cooking water to make a hot tea with lemon and honey, or add salt and pepper and vinegar and use as a soup base. The young plants can be used to make tea, wine or beer. Gather the roots in the spring and cook them for a high starch addition.

Notes: The mature stems have a tough fiber used to make thread and twine, and woven into cloth. Leaves are used as a substitute for rennet. A yellow dye is obtained from the roots.

Garden notes: Yes, you could have stinging nettle in your garden if you are brave enough! Brave to work around the stinging hairs, but also because they can be very aggressive. On the positive side, they are valuable for adding nutrients into the soil after being turned under, and are thought to keep many pests at bay. All you need to get started is a little piece of root.

Caution: Diabetics should monitor their blood sugar if taking nettles. Internal consumption of older plants can cause kidney and urinary tract irritation. An antidote for the sting is nettle plant tea or wash used externally. Other antidotes are dock, plantain, violet, mullein and hound's tongue. The stinging comes from formic acid in the hairs covering the plant. Harvest wearing long sleeves and gloves to avoid skin irritation.

Nettle Infusion

Two cups of nettle leaf infusion can provide all the vitamins and minerals one needs for a day. In a quart jar, cover one cup (one ounce) of dried nettles with very hot water and steep for 4-8 hours. Strain and drink hot or cold. Store in refrigerator up to 24 hours. Use leftover liquid as plant fertilizer and place plant matter in compost.

Valerianaceae Family — Valerians — *Valeriana* spp.

Western valerian habit (top right) and leaves (inset)

Edible valerian habit (top left), leaves (above) and edible valerian flower male (right)

Valerian Family

VALERIANS
Valeriana spp.

EDIBLE VALERIAN
Valeriana edulis

Other names: Tobacco root, tall valerian

WESTERN VALERIAN
V. occidentalis

Other names: Common valerian, small-flower valerian

Description: Edible valerian is a perennial with a large thickened taproot like a carrot. The slender, lance-shaped leaves are mostly basal with almost parallel veination and white edges. Small, yellowish to whitish flowers are mostly unisexual with sexes on separate plants in an open inflorescence; flowering stems up to 2 feet. Found in open dry to moist sites from montane to subalpine. Blooms June to August.

Western valerian is a perennial up to 18-20" tall with whitish to pinkish tubular flowers in dense cymes. Leaves opposite and leaflets broad, pinnately divided. Leaves toward the base are spatulate with a distinct "thumb". Has a rank smell. Found in rich, often moist soils of open areas from montane to subalpine. Blooming very early in Spring, May and June.

Medicinal Uses: The root of **western valerian** is preferred for use as a medicine. It is a sedative and antispasmodic with fewer side effects than synthetic tranquilizers. It stimulates the central nervous system, digestion and the cardiovascular system. A few people find valerian to be more of a stimulant, especially in response to fatigue. It is effective for stress induced anxiety, muscle tension and insomnia. The fresh root tea or tincture will lessen menstrual pain, muscle pain, bronchial spasms, intestinal cramps and various nervous disorders. Studies show these plants are also anti-bacterial, anti-diuretic and liver protective. Many people find the taste and smell of the tea and tincture unpleasant so the dried root can be powdered and taken in capsule form. **Edible valerian** is considered ineffective.

Edible uses: Edible valerian is the only valerian considered "edible." I have read varying reports as to the poisonous state of the raw roots, so if this root is consumed it should always be cooked. My recommendation is to smell the plant first, and if you find it objectionable (as I do), you will probably find the plant to taste similarly. I kind of relate the smell to dirty laundry. Hmmm. Some people find the taste earthy and pleasant. Collecting the roots is best done in spring or summer; fall produces very fibrous and worm-eaten roots. The Native Americans used the roots extensively and usually cooked the large roots in fire pits dug in the ground and lined with stones, then grass, with the roots in the center. The roots could be pounded into flour.

Caution: Pertains to **western valerian**: avoid long term use of the dry root preparation. Constant use over a long period of time can cause dependence, emotional instability and depression. Do not use during pregnancy. Valerian tea should never be boiled. Large doses can cause vomiting, stupor and dizziness. Avoid if taking sleep inducing drugs or alcohol; may enhance their actions. It is believed the roots of **edible valerian** are poisonous when raw.

Violets

Violaceae Family *Viola* spp.

Purple violet

White violet

Yellow violet

Violet Family

VIOLETS

Viola spp.

Description: We have several violet species in the area with three main colors: white (*V. canadensis* var. *rugulosa*), yellow (*V. nuttallii, V. purpurea* var. *venosa*, and *V. praemorsa* var. *linquifolia*), and purple or shades thereof (*V. adunca, V. labradorica*). They are all low growing plants with simple leaves. The unique flowers are 5-petaled with 2 upper, 2 lateral and one lower petal that has a nectar-holding spur at its base. All have distinct lines running to the throat. Most species are found in moist, rich soil, shaded areas from foothills to alpine. Most bloom spring to midsummer.

Medicinal uses: Violet leaves contain varying amounts of laxative, diuretic and expectorant compounds. The yellow flowered species is the most laxative. The plant is used in cancer formulas as a "blood purifier" which aids the liver in eliminating excess waste compounds in the blood. Violets are also mildly astringent and antiseptic as well as mucilaginous and can be useful in treating ulcers, lung, and urinary tract and skin problems. Medicinal teas of the above ground parts have been used to treat bronchitis, asthma, heart palpitations and fevers and as a gargle for sore throats and coughs. Poultices or washes are used externally for tumors, bruises, eczema, rashes, etc. The roots were used by Native Americans to induce vomiting in poisonings. The plant is also used for flower essence and as a homeopathic remedy. Since all violets contain similar properties and uses, I prefer to use the much more available garden varieties, Johnny Jump-ups and pansies. Let their wild cousins be.

Edible uses: For all violets, use only the leaves, stems, and flowers. Roots, rootstalks, fruits and seeds contain toxins. All violet leaves are edible and vary in their flavor and are high in Vitamin A and C. **Canada white violet** (*V. canadensis* var. *rugulosa*) has a pleasant tasting leaf and is my favorite. Use leaves in salads, as trail munchies, potherbs or as a thickener. Fresh flowers are also nice in salads and as trail munchies, candy them, or add them to vinegar, jams and jellies, and syrup for some coloring. To make violet syrup, cover blossoms in boiling water and let it sit. Add sugar in the ratio of 1 cup extract to 2 cups sugar and bring to a boil until it thickens. Fresh flowers can be placed, one in each ice cube tray square, making a drink with colorful ice. The leaves have been fermented to make wine.

Notes: The plant is used in cosmetics. Violet flowers have been used as a substitute for litmus paper. Mashed leaves were burned like incense to ward off disease.

Garden notes: Our lovely natives are related to the Johnny jump-ups and pansies that are so popular in the garden. I have had very good luck transplanting Canada violet in my shade areas. Most native violets have really funky pollination and seed formation with seed-producing flowers that never open, they are self pollinated, and open underground or emerge only after the fruit is matured.

> **Caution:** Consumption of large quantities of the leaves can cause digestive upsets. Consuming the roots, rootstalks, fruit and seed can cause vomiting. Violet leaves can be confused with heart-leaf arnica (*Arnica cordifolia*) before flowering.

Equisetaceae Family — Horsetails — *Equisetum* spp.

Scouring rush stems with the unique white and dark bands (right)

Horsetail fertile stem with strobili at the tip and emerging young vegetative stems (top oval); Horsetail habit (above)

Horsetail Family

HORSETAILS

Equisetum spp. *(Hippochaete* spp.*)*

Other Names: Scouring rush horsetail, field horsetail, meadow horsetail, smooth horsetail, bottlebrush, western horsetail

Description: Both horsetail and scouring rush are perennials with jointed, hollow, vertically rigid stems from underground rootstalks often with small tuber-like swellings. Reproduce from spores in cones, called strobili, at stem tips. **Horsetails** have numerous branches in whorls at the nodes of the green main stem. Field horsetail also produces a brown, unbranched stem containing the strobili at the tip. This brown stem is produced in early spring and fades quickly. **Scouring rush** is a stout single green stem without the fine branching. Both horsetail and scouring rush need moist sites, with the horsetail growing all the way into the alpine.

Medicinal uses: The **horsetails** are astringent. The young stems, prepared as tea, can check bleeding in nosebleeds, wounds, intestines, and heavy menstruation. A strong diuretic for urinary tract and prostate problems, they also tonify the urinary mucous membranes, can control bed wetting and help with skin disorders. Horsetails have been used around the world to treat arthritis and water retention. Their high silica and calcium content make it beneficial for mending fractures and connective tissue and strengthening hair, nails and bones. A decoction, when added to bath water can help slow-healing sprains and fractures and irritable skin problems like eczema. Research continues to evolve on the uses and safety of this plant. A homeopathic remedy is made from the fresh plant and used to treat urinary system complaints. Scouring rush can be used in the same way, but is less soluble so probably less effective.

Edible uses: Both **horsetails** and **scouring rush** rootstalks have been eaten fresh or boiled, but they should only be used in moderation, if at all. With some of the horsetails, the green stems were dried, ground into a powder and used as a mush or thickener. This powder can also be used in teas.

Notes: The horsetails belong to one of the most ancient of land plants with abundant fossils showing their structure, which has not changed much over millions of years except for their size. The ancient horsetails from 300 million years ago were over 50 feet tall! They have silica deposited on their stems which serve them quite nicely as pot scrubbers, silver polish, and sandpaper. The hollow stems make great whistles. Light pink dye is obtained from the stem. The infused stems can produce a fungicide and plant fertilizer.

Caution: Harvest only from clean water sources as the plant can take up chemicals like pesticides, heavy metals, etc. Older plants contain potentially toxic amount of selenium and other alkaloids. Only use young plants. High doses and long-term ingestion is not recommended. Ingesting large amounts of uncooked plant material may also cause a thiamine (vitamin B1 deficiency) in humans and animals.

Parmeliaceae Family — Usnea — *Usnea* spp.

Caution: Some lichens can be poisonous. It is important to correctly identify the species before using.

Usnea thallus

Usnea growing on barn wood (above) and *Usnea* with *Bryoria* on dead conifer branch (right). This is a common combination in the subalpine forests

Parmeliaceae Family

USNEA

Usnea spp.

Other Names: Old man's beard, beard lichen, hair lichen

Description: The *Usnea* are also called hair lichens because of their fine, tufted appearance hanging from trees like a clump of hair. Many genus of lichens can be found on trees, but the Usnea are always light gray-green, with a stout white central cord like an elastic thread when the outer sheath is pulled gently apart.

Medicinal uses: *Usnea* has a long history of use around the world. It contains antibiotic substances that been shown to be more effective than penicillin in inhibiting the growth of many gram positive bacteria like streptococcus and pneumococcus. It is an antimicrobial herb which inhibits or kills invading bacteria without disrupting the body's natural flora. It is used to treat lung diseases like TB, pneumonia, upper respiratory tract infections, sinusitis, bronchitis and whooping cough and for other problems like strep throat, urinary tract infections, and diarrhea. *Usnea* is also effective against viruses including herpes simplex and Epstein Barr and fungi including Candida species. Fungal infections like athlete's foot, dandruff, ringworm and vaginal infections respond to *Usnea*. It also has a stimulating effect on the immune system. Because of its bitter taste, it has a history of being used to treat indigestion. It can be used internally as a tincture or externally as a salve, douche or wash. Preliminary test tube studies have indicated possible anticancer actions. This lichen can be used right off the tree as a first aid antimicrobial compress to prevent bleeding and infections on wounds. Visiting the forest after a windy day often provides enough fallen branches with *Usnea* attached to provide a nice harvest.

Edible uses: *Usnea* itself is not considered edible, but many lichens are. I have tried rock tripe, *Umbilicaria*, by boiling it for 10-15 minutes (I have read that several changes of water make it less bitter). It would pass fine for a wilderness survival food.

Notes: The plants produce a yellow dye. Lichens seldom need a mordant, unlike most other plants.

A Little Lichenology

"Lichens are fungi that have discovered agriculture."
— Lichenologist Trevor Goward

Yes, the lichens we "see" are fungi, which are incapable of making their own food, so these fungi "farm" partners that can photosynthesize: algae and cyanobacteria (blue-green algae). Along with the nutrients from the algal partners, lichen fungi absorb whatever is in the air. What goes in cannot be purged from the lichen, so if there are pollutants in the air or precipitation, these will be found in the lichen. *Usnea lapponica* was one of the lichens used in a biomonitoring study in 1996 to determine where pollution was coming from that was affecting the Mount Zirkel Wilderness Area north of Steamboat Springs. The most likely causes were the metals, nitrogen and sulfur gases the coal-fired power plants in the Yampa Valley emitted. There was a signature sulfur that was linked to the coal burned in these power plants that was mined in local coal mines. This unique sulfur isotope was found in high levels in *Usnea* closest to, and downwind from, the power plants. Lichens are monitored around many cities as kind of a canary in the mine. Some lichens are particularly sensitive to certain pollutants and are degraded when exposed long-term.

Poisonous Plants

Red or white, don't take a bite unless you know what it is. (snowberries, poison ivy berries, baneberries)
If it doesn't smell like an onion, it isn't one! (death camas)
Purple splotches on stems, use caution. (poison hemlock)
If it smells really bad, maybe you should reconsider. (hemlocks, false hellebore)
Wild pea pods are NOT the same as garden peas, do not eat! (peavine, lupines, vetch, locoweed, golden banner, milk vetch)

During plant walks and classes we find ourselves saying these cautions over and over as a way to offer simple tools to prevent nasty experiences. There are many stories of people who have died from misidentification of a poisonous plant. We do our best to prevent that from happening. Not all plants are deadly but some have the potential to cause negative reactions if taken in larger doses, during pregnancy or exposed to sensitive skin. We have tried to highlight potential problems in our **Caution boxes** with each plant. This section of the book covers the plants which are truly those with the most toxic effects. We recommend doing your own research. Use up to date resources. Many older books, especially old herbals, have many inaccuracies based on unverified anecdotal information, different plants or uses or unreliable research. We have come a long way since the days of the early herbalists like Culpepper, so please use modern resources as your primary sources of information. When doing your research watch for chemical constituents that can give you a hint there might be some toxicity: alkaloids, glycosides, alcohols, organic acids, terpenes, saponins, resins and resinoids.

Plants have many systems of self-protection from predators, disease and competition. These can take the form of toxic chemicals, stingers, thorns, irritating hairs, etc. Some plants have mechanisms that favor its distribution and survival. Some birds can tolerate large amount of toxic fruits so it can distribute the seeds through its digestion.

The potency of a plant varies not just from species to species but from plant to plant, parts used, how and when they are harvested, and how prepared.

Some plants are poisonous only at certain stages of growth. Lupines are most poisonous in late summer when the pods are full of toxic seeds. Larkspur is very deadly in early spring and loses it toxicity almost entirely at flowering time. Death camas bulb is very deadly in spring but is less harmful by the time it dries up in July. The roots of poison hemlock contain less poison during the spring than the leaves and stems do, and the roots become toxic later in the season.

Unusual conditions may affect the quantity of poison in plants. Wilting can increase toxic chemicals in some plants, like the leaves of wild cherry and clovers. The skins of our common potato develop dangerous levels of solanine when exposed to the sun and turn green, although you have to eat a large quantity to have negative effects. Normally safe plants can accumulate harmful chemicals from contaminated sites like mine tailings, commercial agricultural fields, septic waste areas and fracking sites. Smoke from burning a poisonous plant may carry toxins.

Poison can be found in different parts of plants. In some cases the whole plant is poisonous like the hemlocks, false hellebore and monkshood. Many roots contain higher amounts of poisonous constituents than other parts. Saps and juices of many plants cause contact dermatitis—buttercups are an example. In showy milkweed, the stem sap is considered toxic. In Pea Family plants, the ripe seeds within the pods contain the most risk. The leaves, stem, bark and seeds of chokecherry contain varying amounts of cyanide producing compounds, with the wilted leaves producing higher amounts. Leaves, twigs and fruits of serviceberries can also contain small amounts of these cyanide chemicals. Phototoxins are constituents that can make the skin extremely sensitive to sunlight through simple contact on the skin or eaten and absorbed into the blood stream. Some members of the Parsley (Apiaceae) and Citrus Families and specifically St. John's Wort have these chemicals. The hairs on stinging nettles inject chemicals into the skin that cause pain and irritation.

Poisons can vary according to season, climate, etc. Drought generally favors the production of poisons in most plants. Some plants become less poisonous with cultivation, like monkshood. As discussed above, the time of year can affect toxicity.

Some people and animals are more susceptible than others. Light skinned people are more likely to be affected by phototoxins than dark skinned individuals. Horses, mules and goats can eat poison ivy with no problem while 85% of humans have a violent skin reaction just touching it and, once exposed, are said to have an increased sensitivity. Pregnant women may react differently to exposure or ingestion. People taking drugs may have different reactions. Allergy producing chemicals in some plants can cause strong reactions in susceptible individuals; skin rashes and pain (poison ivy, nettles, cow parsnip) and runny nose to severe asthma attacks (ragworts, grasses, goldenrods). This is why we constantly recommend a gradual and tested introduction to a plant before using it.

Dosage can determine toxicity. Luckily some plants taste so bad that eating too much is not a problem. Some plants are fine in small amounts but can cause problems if taken over a long period of time and accumulate in different organs of the body. In herbal preparations, an experienced herbalist using toxic plants will make a very dilute remedy and use dosages in single drops.

Preparation can affect toxicity. Many books talk about boiling plants more than once and discarding the water to remove toxins. Some plants must be dried or heated to dissipate toxins—chokecherries are an example. Many plants considered toxic have constituents that can be very helpful when prepared into drugs. Learn these poisonous plants thoroughly so you can explore all the other usable plants in nature. Be sensible, but don't be paranoid.

Poison Ivy Meme

"Leaflets three, let it be; leaflets five, let it thrive.
Berries red, have no dread; berries white, a poisonous sight.
Longer middle stem; stay away from them.
Hairy vine, no friend of mine. (poison ivy vines found in eastern US)
Raggy rope, don't be a dope. (poison ivy vines found in eastern US)
Side leaflets like mittens, will itch like the dickens. (some leaves have a "thumb")
Red leaves in the spring, it's a dangerous thing."

What To Do:

Recommended treatment for poisoning seems to be an evolving science and cannot be covered thoroughly here. If patient is unconscious, call 911. Otherwise, call Poison Control and seek medical advice. In most cases inducing vomiting is no longer recommended. Give a conscious and non-convulsing patient milk to absorb the poison until they can get to a hospital. Keep samples of the suspected plant for positive identification.

Anacardiaceae/Apocynaceae Families — Poisonous Plants

Poison ivy leaves (above inset) and in fall color (oval)

Poison ivy habit

Poison ivy white berries in winter (above right inset)

Spreading dogbane habit

Spreading dogbane leaf and flower

Sumac/Dogbane Family

WESTERN POISON IVY   *Toxicodendron rydbergii* (*Rhus radicans*)

Most folks have heard of poison ivy and should to be able to identify it. Yes we do have it here in our higher elevations of northwest Colorado. The plant is most dangerous in spring and summer when sap is abundant. Fifty percent of the population is allergic to some degree. Painful, irritating redness, itching, swelling and blistering can result from just touching any parts. The oily irritating substance, urushiol, is released with the slightest bruising and can be transmitted on pet fur, exposed clothing or tools. Inhaling the smoke from burning the plant can also cause serious irritation and blistering internally. Washing immediately (within 5 minutes) after exposure, even if mountain water is all that is available, is the best treatment. Be sure to wash all clothing, tools and pets to prevent spreading the oils and recontamination. Some people with severe reactions should consult a doctor. Karen, a frequent sufferer, recommends TecNu.

Description: Woody stems single and erect or vine-like, trailing over the ground and lower vegetation. Alternate leaves divided into 3 egg-shaped, pointed leaflets, glossy and bright green, turning scarlet in fall. Remember; Leaves in three, let it be! Inconspicuous yellow-white flowers in crowded clusters from leaf axils, producing white or yellowish-white berries. Found at increasingly higher elevations (several sites in Steamboat Springs as of this writing), in rocky, moist forested areas or open sites. Blooms May to July.

SPREADING DOGBANE   *Apocynum androsaemifolium*

Of the two species of dogbane in Colorado, the most prominent in our area is spreading dogbane. Both have a long history of use among Native peoples externally as a counter irritant to increase blood flow to the skin. Internally these unpleasantly bitter plants were used as a birth control and to treat coughs, rheumatism, whooping cough, parasites, diarrhea, headaches, convulsions, heart palpitations and more. Early settlers used the roots as a laxative and to induce sweating. Some experienced herbalists still use them today but with great care. Dosage must be highly dilute and controlled due to their actions on the heart and digestive system. The sap has been used to treat warts, although it can blister the skin of some people. The plants also yield a type of latex that can be used like rubber. I think this is an example of where indigenous knowledge of how to safely use some of our more complex plants has been lost to modern users.

Description: An erect, widely branching perennial plant up to 20" tall with often reddish stems with milky sap, forming large colonies from underground rootstalks. Opposite leaves mostly along the stem, drooping, egg-shaped to oblong. Sweet smelling, bell-shaped flowers in few-flowered, showy clusters at the top of the stem, white or pinkish with dark pink veins; 5 fused petals curl back at the tip. Found in open or shaded rocky areas from montane to subalpine. Blooms June to September.

Apiaceae Family — Poisonous Plants

Poison hemlock habit (above), leaf and purple blotched stem (top right)

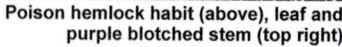

Water hemlock flower (below left), and habit (below right)

Water hemlock leaf

Parsley Family

POISON HEMLOCK Conium maculatum

Though this plant has a history of use by medical doctors in the 18th and 19th centuries it is rarely used today. All parts are poisonous, but the toxins are most abundant in green seeds. The toxins affect the nervous and respiratory systems, causing numbness and paralysis of the arms and chest, leading to suffocation. Poisoning symptoms include nervousness and confusion, weakness, vomiting, diarrhea and weak pulse. It has a history of being used to commit suicide. If ingested, induce vomiting, administer a strong laxative and immediately consult a doctor. Even handling the plants can cause reactions in sensitive people. Always wash your hands afterwards. It is a member of the *Apiaceae* (Parsley) Family of plants with several look-alikes. Learn to ID these plants! If you have this plant growing in your living area it is best to eradicate it. There really is no reason to use either of the Hemlocks with the wealth of other, safer plants available from Mother Nature.

Description: Tall, rank, non-native biennial up to 8 feet tall with a nasty musty smell. Stout stems are purple blotched, and the leaves are lacey and fern-like. White flowers are arranged in an umbel, these umbels blooming in a staggered fashion along the length of the upper stem throughout the summer. Found in disturbed sites from foothills to montane. Blooms June to August.

SPOTTED WATER HEMLOCK Cicuta maculata

Though Native Americans used parts of this plant for medicine their understanding and expertise at using it is beyond what we know today. Do not use this plant; it is not worth the risk. Get to know how to identify this and other members of the *Apiaceae* (Parsley) Family. Some are great medicines, others are deadly. Many Native Americans used the roots of this plant to commit suicide. Water hemlock is considered the mostly deadly plant in North America. Small amounts (2-3 bites for humans) can kill. Even children making pea shooters or whistles from the hollow stems will be poisoned. The poison acts on the central nervous system and can take effect within 15 minutes. Symptoms include stomachache, salivation, nausea, vomiting, diarrhea, difficulty breathing, tremors, and violent convulsions. If ingested, induce vomiting, administer a strong laxative and immediately consult a doctor. If this plant is growing in your living area, it is best to eradicate it, especially if you have kids or pets.

Description: Tall, rank, native perennials up to 4 feet tall with foul-smelling, oily yellow sap and thickened, chambered stem bases and/or roots. Alternate leaves pinnately divided leaves with side veins ending at base of teeth. Stems can have purple spotting. Flowers are white to greenish umbels. Found along ditches, streams, common in urban wet areas, in foothills and montane. Blooms June to August.

> Ancient Greeks were familiar with the deadly properties of poison hemlock. Its juice was given to enemies and criminals. In 399 the philosopher Socrates was convicted of heresy and sentenced to death. He drank hemlock tea creating a paralysis that moved from extremities toward his respiratory system and heart. He had no pain, his mind stayed clear and he was able to talk until the end.

Asteraceae Family — Poisonous Plants

Heart-leaf arnica leaf and flower

Heart-leaf arnica habit and fruit (oval)

Tall ragwort leaf and flower (center) and coarse flowers (below right)

Arrowleaf ragwort habit (far left) and flower (left inset)

Sunflower Family

ARNICAS Arnica spp.

The form of arnica used internally is a popular homeopathic remedy for shock and trauma, bruises and sprains. The entire above ground plant in bloom has been used for centuries externally to make compresses, liniments. massage oils or ointments for bruises, sprains and all types of muscle and joint pain. External applications act as an irritant to increase and stimulate white blood cell activity, improve local blood supply, speed healing, are anti-inflammatory and increase the rate of re-absorption of internal bleeding. Do not use on broken skin or open wounds as it can increase bleeding and introduce potential toxins into the bloodstream. Do not take internally. The plant can cause gastroenteritis. Some people are sensitive to this plant and have reported dermatitis with external use.

Description: There are many species of arnica in our area. Heart-leaf arnica (*A. cordifolia*) is a perennial 6 to 24" tall with mostly basal, heart-shaped leaves, opposite leaves along the stem are velvety, ovate and sessile. 1 to 3 flowers per stem with yellow ray and disk flowers, up to 3" across. Found in conifer or aspen forests from foothills to subalpine. Can bloom once in the spring then again in the fall. It is currently on the "To Watch" list of overharvested plants. Be considerate. Other fairly common species include *A. parryi, A. chamissonis* and *A. mollis*. Consult a field guide for positive identification.

Garden notes: *A. cordifolia* is for sale in very specialized native plant nurseries and can be difficult to get established. This is a high priority plant for gardens to help reduce overharvesting in the wild.

RAGWORTS Senecio spp.

Some *Senecios* contain pyrrolizidine alkaloids which can cause liver damage and be carcinogenic to animals including humans with long term exposure. These chemicals are highest in the flowers and lowest in the roots and leaves but this can vary greatly between species and plants. It appears that our native species do not have high levels of the toxic alkaloids but not much research is being done to verify that. Drying does not destroy the alkaloids. Some species have been used as medicine for menstrual problems, internal hemorrhaging, kidney stones, as a purgative and emetic and as a wash or poultice for wounds. Some sensitive people develop rashes just from handling the plants. Pregnant women should avoid ingestion. It would be best to avoid using this plant altogether. *Senecios* are a host plant for the Field Crescentspot orange butterfly.

Description: The genus *Senecio* is one of the largest in the world, although the genus has recently been further split into *Packera* and *Ligularia*. In Colorado these DYCs (damn yellow composites!) are the most difficult to identify but all have a single neat row of narrow phyllaries enclosing yellow to orangish flowers (often small in relation to plant size), with a pappus of fluffy white hairs. Two common plants, *Senecio triangularis* (arrowleaf ragwort) and *S. serra* (tall ragwort) are tall, leafy-stemmed perennials with clusters of small yellow ray with darker disc flowers. Arrowleaf ragwort has unique triangular shaped leaves and grows in moist areas of the subalpine. Tall ragwort has linear-lanceolate leaves and grows in meadows of the montane and subalpine.

Caprifoliaceae Family — Poisonous Plants

Red elderberry fruit (oval)

Red elderberry flower and leaf (above)

Mountain snowberry fruit

Mountain snowberry leaf and flower

Mountain snowberry habit

220

 Honeysuckle Family

Red Elderberry *Sambucus racemosa* var. *racemosa*

Blue elderberry (*S.nigra* ssp. *cerulea*) is an important herbal medicine in today's commerce but, sadly, our red-berried species is considered toxic and I can find no history of medicinal use. The bark, stems, leaves and roots contain alkaloids that can produce vomiting and diarrhea. The fruits of red elderberry are edible but have been known to cause stomach upset eaten raw. Cooking destroys its toxins. The hollowed pithy stems have been used as whistles and flutes (although this is not recommended as the raw juices are toxic). The branches are favored as drills in fire making.

Description: Shrubs up to 9 feet tall with pithy stems. The opposite, pinnately compound leaves are large and not glaucous and the twigs are covered with warty bumps (lenticels). Small white flowers are arranged in a pyramid-shaped cluster. Fruits are clusters of bright red to almost purple-black berries. Common in forest openings, meadows and old clear-cuts of montane and subalpine. Blue elderberry is not found in our area but could be grown in our gardens.

Garden notes: There has been a plethora of introduced cultivars of elderberry in the nursery trade in the past years. Our native species is seldom offered. Too bad. Native species benefit pollinators, birds and other wildlife more than the new, bigger, fancier varieties.

Mountain Snowberry *Symphoricarpos rotundifolius* (*Symphoricarpos oreophilus*)

The berries of mountain snowberry are slightly toxic if eaten. The symptoms include vomiting, diarrhea, depression and sedation. Other snowberry species have a history of Native American use as eye medicine and cleansing laxative and to detoxify after poisoning. In other words, it is an emetic. Most tribes considered more than 2-3 snowberries poisonous and we should, too.

Description: Heavily branched shrub up to 4 feet tall with peeling older bark. Opposite leaves vary greatly in size and shape, but are generally small and elliptic. Pale pink, pendulous flowers are long tubular bells in few-flowered clusters under the branch. Snow white berries are egg-shaped. Very common, often forming understories, in aspen and mixed oak habitats, as well as canyons and gulches. Blooms May to June.

Garden notes: There is a dizzying variety of snowberries available in nurseries, unfortunately most of them misnamed. Our true native species can be fairly easy to find as long as you know how to identify them. They are tough shrubs, providing good fill in part shade xeric conditions.

Boraginaceae / Fabaceae Families — Poisonous Plants

Western stoneseed habit and flower (oval)

Mountain goldenbanner flower

Mountain goldenbanner habit

Borage / Pea Families

Western Stoneseed *Lithospermum ruderale*

Lithospermum species were once used by many Native peoples throughout North America. The root of our species, Western stoneseed, was used in a decoction as a diuretic for kidney problems and as an astringent for internal bleeding and diarrhea. A poultice of the dried leaf and stems was used for rheumatic pain. The root tea was taken as a contraceptive with efficacy ranging from temporary to permanent sterility. Research has shown that root alcohol extracts have dramatic hormonal effects that eliminate the menstrual cycle in mice and decrease the weight of the thymus and pituitary glands. The plant contains toxic alkaloids. It should not be taken internally. The seeds were used as beads and the roots provide a red dye.

Description: A stout, many-stemmed perennial up to 24" tall with alternate, hairy, dark green linear to narrowly lance-shaped leaves along the stem. Pale greenish-yellow flowers are funnel shaped with 5 spreading lobes in small clusters from axils in upper leaves. Flowers differ in size between plants. Fruits are shiny white to brownish nutlets; as hard as stone (hence the common name). The genus name *Lithospermum* is derived from the Greek *lithos*, 'stone', and *spermum*, 'seed.' Found in oak brush and warm, dry, open sites in montane. Other common names include Western gromwell and puccoon. Blooms April to June.

Pea Family *Fabaceae*

Many members of this family produce pods that resemble the garden pea pods many of us are familiar with and love to feast on. Do not feast on any pea-like pods you find in the wild unless you are absolutely sure of your identification. They are hard to identify and many are poisonous. The good thing about this family is that they help improve our soils by fixing nitrogen.

Mountain Goldenbanner *Thermopsis montana*

This early blooming plant has pods that have enticed children to eat them resulting in sickness and even death. They contain toxic alkaloids which have been known to poison cattle and horses, as well as anagyrine, a chemical causing birth defects. They can cause calf deformities when eaten by pregnant cows. As little as 18 ounces of ripe seed is considered lethal to livestock. Native peoples used this species as a cough medicine, sore eye medicine and witch protection. Enjoy this pretty plant but make sure you don't enjoy it as food!

Description: A showy perennial up to 24" tall forming extensive patches from underground rootstalks. Alternate leaves are divided into 3 leaflets, stipules are almost linear. Bright yellow pea-like flowers are in dense terminal racemes, producing erect hairy seed pods. Found in moist open to shaded sites in montane. Blooms May to June.

Fabaceae Family — Poisonous Plants

American vetch flower

American vetch habit

White flowered peavine habit

Pea Family

American Vetch Vicia americana

There is history of this plant being used by Native Americans. The young shoots were boiled or baked and the seeds and seedpods eaten. Medicinally the leaves were rubbed between the hands and applied to spider bites. A crushed leaf infusion was used in a bath for soreness. A plant infusion was used as an eyewash. A leaf infusion was used by women as a love medicine. Some vetches do have toxic alkaloids which can attack the red blood cells of persons with a certain enzyme deficiency. Others may have chemicals that are toxic to the nervous system. This species may not contain these chemicals but, for me, I would err on the side of caution and avoid it. Vetches, locoweeds and peavines all look really similar to the untrained eye. Use with caution and if you are unsure, don't!

Description: Slender, vine-like perennial trailing or climbing up to 3 feet that form large clumps from underground rhizomes. Pinnately divided leaves have a coiled tendril at the end that wraps around items for support. Reddish to bluish-purple pea-like flowers are in loose clusters from the leaf axils, producing smooth pea-like pods. Common in meadows and open forests in foothills and montane. Blooms May to August.

White-Flowered Peavine Lathyrus lanszwertii var. leucanthus

Numerous species in this genus are known to produce "lathyrism" if large quantities of the seeds are eaten. Lathyrism can cause a nervous system reaction with symptoms ranging from spastic paralysis of the legs and arms to degeneration of the spinal cord. Both humans and livestock poisonings have been reported. Small amounts can provide a nutritious food but if eaten in larger amounts and regularly for long periods the nervous system can be damaged by the neurotoxic amino acids they contain. Great caution is advised!

Description: A slender, usually climbing perennial up to 20" tall with pinnately divided leaves having a tendril at each tip that curls only slightly. White pea-like flowers in clusters at the leaf axils age to a soft rust-orange. Pea-like seeds are smooth. Found in open woods from montane to lower subalpine. Blooms May to July.

V for *Vicia* or Lazy *Lathyrus*??

Take a leaflet, turn it over and notice the veination. If the veins run in a straight line from the midrib out to the edge in a "V" you have a *Vicia*. ("V" for *Vicia*. Get it??)

If the veins are lazy and wander around from the midrib to the leaf edge, you have a "lazy" *Lathyrus*. Easy, huh?

Fabaceae Family   Poisonous Plants

Lambert's locoweed flower

Lambert's locoweed habit

Rocky Mountain milk vetch

Alpine milk vetch

Pea Family

LOCOWEEDS *Oxytropis* spp.

Locoism is the name for the trembling, listlessness, lack of appetite (except craving for more of this plant), impaired vision, paralysis, prostration and death that can occur in as little as one month after an animal first begins eating this plant. Cattle, horses and sheep are susceptible to being poisoned. These plants contain toxic alkaloids and take up selenium from the soil. Do not eat!

Description: Most of the locoweeds are quite showy and variously hairy, some almost woolly. The leaves are pinnate, typically basal. The pea-like flowers have the keel petals narrowed to a point like an abrupt beak. Most are found in dry, poor soil sites from the foothills up to the alpine, with most of our species found at higher elevations. **Lambert's locoweed** (*O. lambertii*) is probably the most spectacular with bright magenta displays in sagebrush in mid-summer. It is also considered one of the most poisonous plants in the Western U.S. A few other natives are *O. deflexa*, *O. podocarpa* and *O. sericea*.

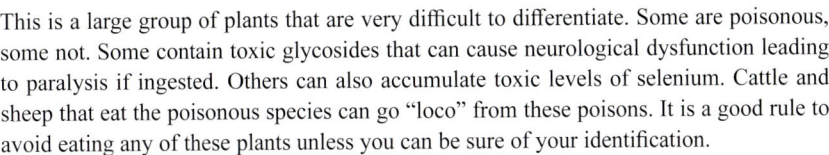

MILK VETCHES *Astragalus* spp.

This is a large group of plants that are very difficult to differentiate. Some are poisonous, some not. Some contain toxic glycosides that can cause neurological dysfunction leading to paralysis if ingested. Others can also accumulate toxic levels of selenium. Cattle and sheep that eat the poisonous species can go "loco" from these poisons. It is a good rule to avoid eating any of these plants unless you can be sure of your identification.

Description: The milk vetches are perennial herbs that can seem initially overwhelming until one realizes that typically only a few species occur in an area. Using habitat will make the task of identification easier. Many milk vetches are quite showy, especially in the spring. The stamens are united by their filaments (10 together, or 9 united and 1 separate), and the lower-joined petals (the keel) is canoe-shaped. The pods are not joined between the seeds. Here are two of the more common milk vetches: **Rocky Mountain milk vetch** (*A. scopulorum*) is yellowish or cream, lanky, up to 2 feet tall (often sprawling), seed pods compressed deeply along the sides, common in scrublands, woodlands from foothills to montane in June to July. **Alpine milk vetch** (*A. alpinus*) is small in stature but forms larger clumps, flowers white and purple, seed pods with dark hairs, found along streams and wet meadows of montane to timberline.

Fabaceae Family — Poisonous Plants

Stemless dwarf lupine habit (left) and palmately divided leaf (round inset)

Silver lupine habit

Pea Family

LUPINES *Lupinus* spp.

Not all lupines are toxic, but it is very difficult to distinguish between nontoxic and poisonous species. Seeds and young leaves contain toxic alkaloids and are the most dangerous parts. Symptoms may include weak pulse and nervousness, depression, labored breathing, convulsions, and paralysis. Lupines have caused birth defects of bone in cattle, termed "crooked calf disease," in which calves are born with bent legs or arched backs. Bone deformity was reported in a human baby whose mother, during pregnancy, drank milk from goats feeding on Lupinus latifolius. Non-poisonous species were said to be eaten by native tribes. They also obtained a greenish-yellow dye from the plant. In Europe, a lupine species is cultivated for its seeds, but these must be soaked and boiled before eating.

Description: Lupines are another promiscuous genus, forming hybrids that can have you pulling your hair out. All lupine in our area have palmately divided leaves with many species covered in hairs. Terminal racemes or spikes with pea-like blue, purple, pinkish flowers to all shades in between, often with white highlights. The pea-like pods tend to be hairy. Our most common species, **silver lupine** (*L. argenteus*) forms clumps up to 30" tall with light to dark purple flowers. Found in dry montane and subalpine meadows. Blooms May to July. A cute little one is **stemless dwarf lupine** (*L. caespitosus* var. *caespitosus*). Rarely over 8" tall, with short racemes of blue flowers tucked into very hairy leaves. Found on sagebrush, gravel, and open meadows in montane. Blooms May to June.

Nitrogen Fixers

Nitrogen is one of the most important plant nutrients. Nitrogen-fixers are soil building plants that harbor bacteria or fungi among their roots and extract nitrogen from the air and convert it to a plant-available form. A vast majority of nitrogen-fixing plants act as pioneer species in succession and do well in open, disturbed sites. They can colonize low-fertility sites because they provide their own fertilizer. Many also can perform other functions in the landscape; some are edible, others make excellent ground covers or ornamentals. Most nitrogen-fixers are also generalist nectary plants and many provide habitat for parasitic wasps.

Incorporating these plants in our gardens and landscapes can slash the need for fertilizer and at the same time pour more organic matter in the living soil than any synthetic fertilizer ever could. In a new garden or poor soil, having 25% nitrogen fixing plants is not too many. Landscaping with nitrogen fixing trees and shrubs is beneficial and provides more long-term nutrient feeding. Most members of the Bean Family (*Fabaceae*) fix nitrogen and can be used as cover crops, annual or perennial garden additions.

Other useful nitrogen-fixers for our landscapes are: alder, buffaloberry, cattail, chamomile, chicory, chives, columbine, comfrey, common milkweed, Siberian pea shrub, elkbrush, and mountain mahogany.

Helleboraceae Family · Poisonous Plants

Monkshood flower and leaf (oval)

Monkshood flowering habit

Baneberry flowering habit

Baneberry red fruit, leaf and white fruit (oval)

Hellebore Family

Monkshood *Aconitum columbianum*

All *Aconitum* species are considered poisonous and rarely prescribed for internal use. There is history of use of A. napellus in Europe in small doses to treat certain infectious diseases, heart failure and fevers. Externally, it has been added to salves for its pain killing action on neuralgia and rheumatism. There is a homeopathic remedy used for sciatica, neuralgia, chills and other conditions. The therapeutic internal dose is very close to the toxic dose causing numbness, tingling, tightening of the throat, nausea and vomiting, respiratory depression, convulsions and coma. The sap causes numbness and tingling on the skin. Some human deaths have been reported and it is considered virulently poisonous to livestock both before and after flowering. This is a deadly poison in the wrong dosage. Just appreciate its power and beauty and leave it be.

Description: Herbaceous perennial 2 to 5 feet tall with palmately lobed leaves alternate along the length of the stem. Flowers in a loose terminal spike; deep purplish blue upper sepal forms a hood over 2 parallel side wings and 3 small lower petals. The whole blossom is flattened sideways. Found in moist meadows from montane to subalpine. Blooms June to August.

Garden notes: Monkshood is a stunning, majestic addition to a garden with rich soil and moister conditions. I have grown it in full sun with plenty of moisture. Most of the plants available in the nursery trade are hybrids.

Western Baneberry *Actaea rubra*

Native peoples used the root and foliage of this plant to treat menstrual and postpartum problems, colds, coughs and stomach pains, rheumatism and syphilis and externally for boils and wounds. Some experienced herbalists today use the leaf or root in a tincture in small doses as a remedy for various women's complaints. I recommend avoiding use of this plant unless you are working with an experienced practitioner and are well versed in safe preparation and dosage practices. The leaves of baneberry can look very similar to sweet anise leaves. Smell them! If the characteristic sweet anise smell is absent, do not use it. The entire plant is considered poisonous, especially the roots and berries, causing cardiac arrest in large quantities. As few as 2 berries can cause severe cramps, headaches, vomiting, bloody diarrhea and/or dizziness. Do not use in pregnancy or chronic low blood pressure.

Description: Leafy perennial 1–3 feet tall with pinnately divided leaves, coarsely toothed. Flowers are white, very fine rounded clusters, producing glossy white or red berries in rounded to spiked clusters. The white-berried form has a black dot on the tip, making it look like a doll's eye, hence one of its common names. Found in moist, rich soil areas of aspen and open forests from montane to subalpine. Blooms June to July.

Garden notes: The fine, fern-like leaves of baneberry create a beautiful filler in shady, moist gardens. They can be difficult to find in nurseries. Children are very attracted to the pretty berries (red and white) because they look like candy. Fortunately the berries are very bitter. Do not grow these around residential areas.

Helleboraceae Family — Poisonous Plants

Colorado columbine flower, leaf and habit (above).
Western red columbine (inset)

232

Hellebore Family

COLUMBINES Aquilegia spp.

COLORADO BLUE COLUMBINE
A. coerulea (A. caerulea)

WESTERN RED COLUMBINE *A. elegantula*

Though not common today, there is history of using some columbine species to treat conditions ranging from heart palpitations to uterine bleeding. Although no records of toxicity have been seen for our local species, all columbines are probably somewhat poisonous, especially the seeds and roots. These are such beautiful wildflowers, maybe it is best to appreciate them for that beauty and leave them be. The Colorado columbine is our state flower and illegal to harvest. The flowers are sweet and probably safe to eat in small quantities.

Description: Both species named above are slender-stemmed perennials with leaves two to three times ternately compound; the segments roundish lobed. Petals are smaller than sepals and have spurs that protrude from the back of the flower. Sepals are large and petal-like. Columbines are renowned for hybridizing, and even these two species growing close enough together create fascinating hybrids. **Colorado blue columbine** is up to 3 feet tall. Habitat shapes the size and form of the plant. Flowers are many shades of blue, and sometimes pink and white. Found in moist montane to subalpine woods, especially aspen, into alpine rock fields. **Western red columbine** is up to 16" tall, slender, with red and yellow drooping flowers. Very uncommon in our area. Found in dry, shady montane slopes and embankments and in subalpine woods.

Garden notes: The columbines are one of the most highly utilized flowers in our landscapes because they are adaptive to growing conditions and seem to bloom forever. I love their delicate blooms floating in the garden, and the dizzying varieties of colors are sure to mesh with your palette. Because of their promiscuousness, the color of plant you buy will undoubtedly change over the years as the prolific seedlings mature and flower. Surprise! Where did that color come from? Columbines are important pollinator plants for butterflies, hawkmoths, native bees, bumblebees and hummingbirds. Other insects "steal" the nectar by biting the ends of the spurs, thereby cheating the poor columbine out of a transfer of pollen.

National Flower?

"I wonder where Uncle Sam could find a more beautiful or emblematic national flower than the columbine. Its range is general throughout the United States, the name of columbine as well as Columbia is derived from *colomba*, a dove, and is significant of our love for peace. Its generic name, *Aquilegia,* is without doubt derived from the Latin, aquila, an eagle. . .the long spurs suggest horns of plenty while in another position they resemble a liberty cap, and if one takes into consideration the Rocky Mountain variety, *Aquilegia caerulea*, which appears in blue as well as white, we have the national colors-red, white, and blue."

– Arthur Craig Quick (1939)

Helleboraceae Family — Poisonous Plants

Nuttall's larkspur flower

Nuttall's larkspur habit (above)

Subalpine larkspur habit (right) and leaf (circle)

Hellebore Family

LARKSPURS Delphinium spp.
NUTTAL'S LARKSPUR Delphinium nuttallianum (D. nelsonii)
SIERRA LARKSPUR D. glaucum
MOUNTAIN LARKSPUR D. ramosum
SUBALPINE LARKSPUR D. barbeyi

All *Delphinium*s, especially when young, are considered toxic containing many toxic alkaloids. Ingestion of any parts may cause nervous symptoms, nausea, vomiting, depression, weak pulse, difficulty breathing or death in large doses. The seeds have some history of use externally to kill lice and parasites. They are responsible for a significant loss of livestock by poisoning, although sheep and wildlife seem to be immune. Plants seem to have lower toxicity as they age, although seeds are very poisonous. The flowers produce a blue dye and ink. Powdered larkspur petals were used by Navajo medicine men as a sacrifice to the south; the color blue is sacred to the south. The pollen was also used in ceremonies. Larkspurs are pollinated by butterflies and bumble bees. In spite of their toxicity, some species are used as food plants by the larvae of some *Lepidoptera* (butterfly) species.

Description: Most of our delphiniums are taller perennials up to 5 feet tall. Alternate leaves are palmately divided, some plants with mostly basal leaves, most species leafy stemmed. Flowers are irregular with showy and petal-like sepals and smaller, and sometimes hidden, petals. Flowers are predominantly blue to purple with white sometimes integrated. **Nuttall's larkspur**, also called dwarf larkspur, is up to 18 inches with most basal deeply palmately leaves divided with segments further divided. Very common in open woods and meadows of foothills and montane. Blooms May then dies back to root. **Sierra larkspur** has narrow flowers, the upper sepal and its spur forming an almost straight line, and the flowers are usually a beautiful light blue and white. Found in the aspens in drier areas than the other taller larkspur. If you see a tall, robust dark purple larkspur often forming large patches in moist meadows, you have **Subalpine larkspur**. **Sierra larkspur** hybridizes extensively with **subalpine larkspur** in Utah and Colorado to the extent that hybrids are more common in many areas than individuals of either parental stock.

Garden notes: Of course, Delphiniums have been used since the first formal gardens were envisioned because of their stately habit and colors, as well as longer bloom time. Finding the native species is challenging, but they do transplant well.

The Dolphin

The name *Delphinium* is derived from the Latin word *delphinus*, dolphin, referring to the buds looking somewhat like a leaping dolphin. Larkspur is the July birth flower and conveys a feeling of lightness and levity according to the language of flowers.

Iridaceae Family • Poisonous Plants

Rocky Mountain iris habit, fruit (oval upper right) and flower (oval above)

N ⚠   Iris Family

Rocky Mountain Iris *Iris missouriensis*

At one time this plant was used to treat syphilis, but only because it was slightly less nasty and a little less toxic than treatments of the time like mercury and arsenic. The fresh root is toxic and should not be used internally. The root resin irisin can cause stomach upset, irritate the liver, pancreas and gall bladder and cause vomiting, diarrhea and difficulty breathing. Native people did use this plant externally for skin problems, venereal sores and rheumatic joints. A ripe seed paste was used as a dressing on burns. Michael Moore has a long discussion of the history and use of this plant if you are really interested, but unless you are an experienced practitioner, just use the information for curiosity and knowledge. Internal use should be avoided especially in those with acute liver, pancreatic, gall bladder or GI disease and during pregnancy. Native Americans have soaked the roots in animal bile using the juice to poison their arrows.

Description: This wild iris looks similar to our garden varieties with sword-shaped leaves from thickened, spreading rhizomes, and the typical iris-shaped flowers. Our native is pale to deep blue with 3 yellow streaked sepals bending down, and 3 erect petals, hence its other common name Western blue flag. It can be very common in overgrazed wet meadows, as well as along streams from montane to subalpine. Blooms May to early July.

Garden notes: Our wild iris is an easy to grow, long-lived and beautiful addition to any wet areas of the landscape. I have placed them at the edges of bog gardens under roof overhangs and they love it. The yellow patch and dark veination on the petals signal pollinators to where the goodies are, like a pollinator landing strip.

The Fleur-de-lis Symbol

The goddess Iris was considered a messenger between humans and the gods atop Mount Olympus. Wherever she went, a rainbow followed her. Whenever the ancient Greeks saw a rainbow in the sky, it was a sign that Iris was delivering a message to someone. Thus, iris came to mean "rainbow" and, used as the generic name of these plants, reflects the variety of colors sported by many species. One of the goddess Iris's duties was to guide the souls of dead women to the afterworld, and so Greeks often planted the flowers next to graves.

Egyptian kings considered the iris a symbol of power and majesty and used the stylized iris design on their scepters and placed it on the brow of the Sphinx, believing its three major petals to be symbols of faith, wisdom, and valor.

The fleur-de-lis symbol that appears on many royal European coat of arms and flags is based on a stylized iris. Probably *Iris pseudacorus*, a yellow species that grows prolifically in northern France and according to legend, was adopted by the 6th century King Clovis of the Franks, as his family's badge. The symbol came to represent royalty. Some theories say the fleur-de-lis is a lily and others, a lotus. Our blue flag iris is sometimes called the American fleur-de-lis. Some authorities claim the flowers are commonly called flags because King Clovis, King Louis VII of France and other European rulers used the design on flags and banners.

Nashville, Tennessee is considered the Iris City by flower fans and is the official state cultivated flower. Wild blue flag, *Iris versicolor Linné*, is the official provincial flower of Quebec.

- The Secrets of Wildflowers

Melanthiaceae Family — Poisonous Plants

Meadow death camas habit and flower (inset)

Mountain death camas habit and flower (inset)

Colorado false hellebore flower close (far left), young plant just coming up (center oval) and habit (below)

False Hellebore Family

Death Camases *Zigadenus* spp.

Mountain Death Camas *Zigadenus elegans* ssp. *elegans*
(Anticlea elegans var. elegans)

Meadow Death Camas *Zigadenus venenosus*
(Toxicoscordion venenosum var. venenosum)

All parts of these plants contain poisonous alkaloids. Bulbs and new foliage are the most poisonous. Human poisonings result from confusing the young plants with onions. The toxic chemicals are said to be more potent than strychnine, just 2 bulbs can be fatal. Symptoms include dry, burning mouth, thirst, headaches, dizziness, severe vomiting, cardiac irregularities, loss of muscle control, and in severe cases coma and death. They are a serious danger to sheep and cattle because in early spring, it is one of the few green plants available for grazing. **Smell each collected plant. If you harvest a plant you think is an onion and it doesn't smell like an onion, don't eat it!!!**

Description: Differentiate these two plants by size, flower structure, but, most of all, habitat. Both have narrow, grass-like leaves, mostly basal, and white flowers arranged in racemes. **Mountain death camas** flowers are nodding in bud and erect when open; stamens are collected in a funnel near the center, and the tepals have beautiful heart-shaped glands at their base. Fairly common in subalpine meadows and lower moist tundra. Blooms June to August. **Meadow death camas** is a shorter and typically smaller plant, the flowers not nodding in bud, with oval shaped green glands at the base of each tepal, and the stamens spread away from the ovary across the tepals. Found in lower meadows and sagebrush areas of foothills and montane.

False Hellebore *Veratrum californicum* var. *californicum* (*Veratrum tenuipetalum*)

All false hellebores are extremely toxic. They also have a history of use in minute doses as an anticonvulsive, analgesic, heart sedative, etc. The pharmaceutical industry uses some species in drugs to slow the heart beat and lower the blood pressure. I can find no record of use for our particular species. Because of the plants actions on the heart and respiratory system, ingestion can cause nausea, blurred vision, vomiting, diarrhea, severe slowing of the heart, stomach pain, paralysis and even death. The roots are considered the most toxic part of the plant and there are reports of stomach pain just after drinking water from where this plant is growing. Livestock have been poisoned by these plants. Powdered, the plant is used as an insecticide. Native peoples used it as arrow poison. When young it is often hard to differentiate this plant from plants in the Lily Family like false Solomon's seal, glacier lily and others. This is an example of why accurate identification of plants is so critical. Be sure you know your plants and leave this one alone!!!

Description: Tall (up to 7 feet) leafy plants, looking somewhat like broad-leaved corn stalks, hence a common name of corn lily. Underground black rhizomes can often form huge patches. When first emerging in spring, the rounded clump looks like a head of cabbage (do not be fooled!). Growing up to 2" a day, the stalk may produce a large flower stalk (usually they only produce flowers every 8-10 or so years) of hundreds of star-shaped greenish-white flowers with six tepals. Very abundant in moist meadows of upper montane and subalpine. Blooms July to August.

Orobanchaceae Family — **Poisonous Plants**

Paintbrush flower (close-up above) is actually the bracts (modified leaf) surrounding an inconspicuous tube of petals with upper and lower lobes

Paintbrushes in their many showy colors

Broomrape Family

PAINTBRUSHES   *Castilleja* spp.

These lovely plants are so interesting and should be enjoyed for their uniqueness and beauty, not edible or medicinal properties. Native peoples used the whole plant to treat skin and kidney diseases, menstrual problems, as a contraceptive, blood purifier and for venereal disease. They also ate the flowers and sucked on them for nectar. Paintbrushes accumulate selenium, possibly as a defense mechanism against predators. Soils in the western states are high in selenium. Selenium can be toxic and is responsible for livestock birth defects and death. Therefore it is not advisable to use our paintbrushes internally unless you confirm from your local agricultural authorities that your soils are not high in selenium, and then use only in small amounts.

This genus provides nectar for the broad-tailed hummingbird, is pollinated by butterflies, bumblebees and native bees and is a host plant for Fulvia Checkerspot butterfly. Paintbrushes have co-evolved with their main pollinator, the hummingbird, which certainly has to do with the predominantly red color of the genus. Long-tongued insects such as moths and butterflies seem to be attracted to the white and yellow species.

Description: All paintbrushes are showy, feathery spikes of colors from white, yellow, orange to pink, red, purple and all hues in between. The color is actually the bracts (modified leaf) surrounding an inconspicuous tube of petals with upper and lower lobes. Leaves are alternate, few (on most species) and narrow, sometimes lobed. Because the paintbrush can hybridize readily where they grow together, it is sometimes difficult to determine exact species. For example, one of our most common paintbrushes, scarlet paintbrush (*C. miniata* ssp. *miniata*) is typically a scarlet red, unless it mixes with splitleaf paintbrush (*C. rhexifolia*) where orange, pink and many shades therein occur. FUN! Here are other common species of our area; Wyoming paintbrush (*C. linariifolia*) the tallest at 24", western yellow paintbrush (*C. occidentalis*), sulphur paintbrush (*C. sulphurea*).

Garden notes: Many paintbrushes are hemi-parasitic on sagebrush (*Artemisia*). Where sagebrush is lacking, paintbrush parasitize other plants such as grasses, lupine, and currants. Try as you might to grow these lovely plants in your garden, unless you are growing them with their host plant, they will wither and die. Many reputable nurseries are now selling paintbrush grown in the same pot with their host. Good luck!

Indian Paintbrush Legend

This flower inspired the legend of an Indian brave who tried to paint the sunset with his war paints. Frustrated because he could not match the beauty of nature, he asked for guidance from the Great Spirit. The Great Spirit gave him paintbrushes with all the colors he desired. He painted his masterpieces with these paintbrushes, then left them in the fields. According to the legend, the colorful flowers that we see today sprouted from these colorful paintbrushes.

Rocky Mountain Plants (Family Field Guides, Volume 2)

Ranunculaceae Family — Poisonous Plants

Marsh buttercup habit (below)

Sagebrush buttercup habit

Ugly buttercup flower

Ugly buttercup habit

Typical buttercup flower

Pasqueflower fruiting habit

Pasqueflower leaf and fruit

Pasqueflower flower (oval)

Buttercup Family

BUTTERCUPS   *Ranunculus spp.*

All buttercup species are thought to be toxic to some degree. The sap contains protoanemonin, an acrid chemical known to cause skin irritation and blistering when fresh. Symptoms from ingestion include burning of the mouth, abdominal pain, vomiting and bloody diarrhea. Drying and boiling renders the protoanemonins harmless to humans but doesn't improve the taste enough to make it worth the effort. The poisonous roots have been eaten by children with fatal results. Native peoples used buttercup species as external poultices for various skin problems. Muscular aches, colds and other respiratory ailments, and general, unspecified illnesses were also treated with buttercups. A number were used for stimulation and "revival" of unconscious persons. The juice from the flower was used as arrow poison. Historically several of the species have been cooked or dried and eaten as survival food. Livestock are poisoned by grazing on large amounts of the fresh plant, but these cases are rare due to the acrid taste. It is said that beggars in Europe applied the juice of buttercups to their feet, producing blisters to gain deeper sympathies.

Description: The buttercup species typically have 5-8 shiny yellow petals (some lack petals), sepals that differ from the petals, leaves can be divided or simple, and most are found in moist sites (or a dry site during a moist time such as spring melt). A few of our most common buttercups include; marsh buttercup (*R. alismifolius* var. *montanus*), sagebrush buttercup (*R. glaberrimus* var. *ellipticus*) and ugly buttercup (*R. inamoenus*).

PASQUEFLOWER   *Pulsatilla patens* subsp. *multifida*
(*Anemone patens* var. *multifida*)

This plant has been used externally by native people as a counter-irritant for bruises, sore joints and muscles and as a wash to kill lice and fleas. External application can cause irritation and blistering. It can also cause dizziness, nausea, miscarriage, damage the kidneys and lower the heart rate and blood pressure to dangerous levels. Some herbalists use alcohol extracts in minute doses for specific reproductive and nervous system complaints. Homeopathic remedies are used for pain, menstrual, lung and eye problems. The cautions about using this plant are many and it should not be taken internally unless under skilled practitioner supervision. Sheep have been poisoned by eating this plant.

Description: Perennial up to 10" tall forming nice clumps of mostly basal 1-3 times divided (usually in 3's) hairy leaves of slender leaflets, and stem leaves smaller, stalkless, in a whorl near the middle of the stem. Leaves often emerge after the flower. Solitary, cup-shaped purple or lavender flowers with 5-7 tepals, hairy on the outside. Seeds are feathery achenes in fluffy heads. Follow snow melt in dry, rocky open areas from sagebrush lowlands to timberline. Blooms April to August (depending on elevation).

Garden notes: I love this beauty in a garden!! Just as you think winter will never end, here comes a little fuzzball of foliage, followed by a fuzz of a bud, and finally a delicate lavender flower. These are tough perennials, forming large clumps. The true native is tough to find, but there are many other species and cultivars on the market to enjoy.

Ranunculaceae Family — Poisonous Plants

Western blue virginsbower (above)

Sugarbowls (left)

Windflower habit (left), flower (oval) and fruit (right)

Buttercup Family

Western Blue Virginsbower Clematis occidentalis
var. occidentalis (Atragene occidentalis)

Sugarbowls Clematis hirsutissima var. hirsutissima
(Coriflora hirsutissima)

The fresh sap of many Clematis spp. contains protoanemonin in variable amounts. Fresh sap can be highly irritating to the skin or mucous membranes, causing irritation, blistering and internal bleeding. Handling the leaves may cause dermatitis in susceptible individuals. The toxic principle is dissipated by heat or drying. There is history of use of these species externally as counter-irritants for sores and achy limbs and wash for scabs and eczema. Michael Moore has used many species as an internal treatment for headaches. This use should be reserved for advanced practitioners. The seed floss of all species can be use as fire starting tinder and insulation in shoes, etc.

Description: **Western blue virginsbower** is a perennial vine with stems up to 4 feet long that clamber up trees and over bushes or along the ground. Opposite leaves are compound with 3 entire to barely toothed leaflets. Lavender flowers arise singly, nodding from leaf axils, with 4 lavender-blue petal-like sepals. The showy seeds are long feathery achenes in fluffy clusters. Found in woods and forest edges from montane to subalpine. Blooms May to July. **Sugarbowls** can be dense upright clumps up to 24" tall with pinnately divided leaves further divided into linear sections, leaves somewhat hairy. Flowers are unique; pendulous flowers with 4 united leathery tepals that curl back at the tips revealing dark-purple brown on the inside, with paler colors on the outside. The nodding heads turn upright when the fluffy, feather-like seeds ripen, hence hairy clematis as another common name. Found in open meadows and woods in montane. Blooms May to June.

Garden notes: Gather a hand full of the fluffy blue clematis seeds as soon as they separate easily from the receptacle and sow them in a shady, rich soil area of your garden. Like all the clematis, they grow best with their head in the sun and their feet in the shade. Sugarbowls are spectacular in the dry garden.

Windflower Anemone multifida var. multifida

Toxicity of the anemones varies with the species and habitat and information on specific species is limited. But I did find that all parts are considered poisonous when fresh and toxic only if eaten in large quantities. Symptoms include inflammation and blistering upon skin contact with fresh sap. After ingestion, irritation of the mouth, vomiting and diarrhea occur. There is a history of use of the leaves as a counter-irritant for bruises and sore muscles. There is a European superstition that says that the wind blowing over anemones is poisoned. It is highly recommended to hold your breath while running past a field of these flowers.

Description: A slender perennial up to 8" tall with a whorl of somewhat to very hairy stem leaves as well as a few basal leaves; all leaves are palmately divided and further cleft and toothed. Rose colored flowers (with tepals) are single on long naked stems. Found in dry forest openings and meadows of montane and subalpine. Blooms May to July.

Garden notes: Yes, they can be rather small and inconspicuous, but the windflowers have a delicate beauty I love in the rock garden. With a little moisture, it will grow into a nice little clump. Flowers in the Ranunculus Family offer exposed nectar to pollinators and are especially favored by hoverflies as a food source.

Appendix

MEDICINAL PREPARATIONS

Most herbal remedies were originally taken internally as teas and used externally as a poultices, washes, ointments or powders. In more recent times alcohol extracts have become popular. There are many great herbal resources available with recipes (see **Resources**). What I offer here is just an overview of the preparation techniques.

Drying

We are lucky in Colorado to have low humidity. Herb drying is usually quick and mold free. Air drying is a great way to store the major part of summer's herbal bounty. It reduces large quantities of herbs to amounts that are easy to store and use (10 pounds of fresh herbs equals about 1 pound of dried herbs). The secret to successful drying begins with harvesting the herbs at their peak. See **Gathering Guidelines and Ethics**.

There are basically two drying methods: bunching and tray or screen. Soft parts (leaves and flowers) harvested with the stems like Porter's lovage or sweet anise can be bundled into thumb-thick bundles, fastened with a string or rubber band and hung in a cool, airy area out of direct sunlight (light destroys the herbs' valuable constituents and color) till crispy dry. The leaves are then stripped off the stems, labeled and stored in air tight containers out of direct sunlight. I do not like to use plastic bags, but paper bags are fine. Keeping the herb as whole as possible will extend its shelf life.

Loose leaves, seeds and fruits can be dried on newspaper, cardboard or screens covered with cloth to prevent contact with the metal. I think the best drying trays are the shallow cardboard boxes that come from liquor stores when you buy a case of beer. The cardboard helps absorb any moisture, they are stackable, and I can write on the box the plant name and date harvested. I speak from experience – if you don't label the drying herbs in some manner, you will be hard-pressed to identify them when they have dried and shriveled unless they have a distinctive smell. Store as above. Leaves and flowers don't retain their active properties very long and should be replaced each season.

Harder plant tissues like inner bark and roots are processed differently. When harvesting for inner bark I prefer to use branches rather than take from the tree. This avoids the danger of girdling and really harming the tree. Strip the outer bark till you find the green inner layer, which is the medicinally active part and chop into ½"- 1" pieces. Roots usually need to be washed free of clinging dirt then chopped in ½" - 1" pieces. These can be dried like the leaves and seeds on newspaper, cardboard or screens out of direct sunlight till thoroughly dried. This may take several weeks. Do not put into an air-tight container until you are sure the plant materials are dry or they can mold. If not sure of dryness, store in open paper bags. Roots and bark can keep for several years. Store all dried herbs out of direct sunlight and heat, both which destroy the active constituents.

Herbal Teas, Infusions and Decoctions

Making herbal teas may be the easiest of all herbal remedies. Teas have many of the nutrients but may not be as strong medicinally as infusions and decoctions. They contain the water-soluble components of the herb and should be used immediately or within 1-2 days if refrigerated. Covering the cup or container while the herbs are steeping is important in all methods to prevent loss of valuable medicinal components, especially the easily

Appendix

dissipated essential oils of aromatic herbs. Teas can be drunk hot or cold. Strong teas can also be used externally for compresses, washes and in the bath.

To make an herbal beverage tea:

Bring good, clean water to almost boiling
Place 1 teaspoon dried or 2 teaspoons fresh herb in a cup (or tea ball or reusable muslin bag)
Pour heated water over the herb and cover immediately
Allow to sit and steep for 5-10 minutes.
Strain and enjoy

Infusions and decoctions are medicinal-strength teas. The part of the plant you are using determines the method of preparation. Infusions are prepared from softer plant parts like leaves, stems and flowers.

To make an infusion:

Bring good, clean water to just boiling
Place 1 teaspoon dried herb or 2 teaspoons fresh herb in a cup (or tea ball or reusable muslin bag)
Pour heated water over the herb and cover immediately
Steep for 15-30 minutes
Strain and enjoy

Sun infusions are fun and medicinal

Cover the herbs with warm water in a glass jar with a tight fitting lid
Steep in the sun for several hours
Strain and enjoy

Similar to an infusion, a decoction is necessary when you are making remedies from tough plant materials, such as roots, bark or seeds.

To make a decoction:

Place thinly chopped plant material into a glass or stainless steel saucepan and add cold water
Bring the water and herbs to a boil
Reduce heat and simmer for 15-20 minutes
Strain and enjoy.

Compresses

A compress is used to apply a strong herb solution directly to the skin. Soak a clean cloth in an infusion or decoction. Apply the cloth to the skin and let it sit there till cooled. The process can be repeated as needed. This method is used to decrease inflammation, prevent infection, reduce pain and can be combined with massage for relaxation.

Poultices

A poultice is similar to a compress but uses the whole plant rather than the extract. Mash or bruise the herb and apply directly onto the skin. Hold in place with a gauze bandage. These are used to treat inflamed, painful areas or a wound.

Appendix

Baths

Herbs can be used in the bath by adding a strong infusion or decoction to the bath water or by putting herbs in a muslin bag or washcloth and adding to the water. Soak for at least 20 minutes or longer. Use ¼ to ½ cup herb per bath.

Foot soaks can be made the same way in a large wash basin. Steep for 10 minutes; add cool water to comfort level and soak till the water is cool.

Herbal Oils

Herbal oils are made by extracting the plant's oil-soluble constituents and volatile oils. Any vegetable oil will do, yet olive, almond, sesame and sunflower oils are the best. I always use organic oils. Infused oils can be used for cooking, making salves, creams, massage and body oils. Herbal oils can be infused by two methods: cold infusion and hot infusion.

To make a cold infused oil:

Fill a large clean jar with selected herb flowers or leaves. Use dried plant material or fresh herbs that have been wilted briefly to reduce moisture content and prevent mold formation. Make sure the herb is completely covered with oil. If after a few hours the herb has absorbed all the oil, add more oil. Initially, some plant materials release gases so a tight lid is not recommended. Cover with a clean piece of cotton cloth and secure with a rubber band or canning jar lid ring or loosely screw on a lid. Place jar outside in the sun (my preference because I love the energy of the sun and moon in my preparations) or in a warm place inside like a sunny windowsill, wood stove, or furnace for 2 weeks. Stir or shake occasionally. Strain and store in a clean dark glass bottle and label.

Hot infused oil is a quicker method:

Place 2 cups oil to 1 cup of dried herb or 2 cups fresh herb, in a glass bowl over a pot of simmering water or in a double boiler and heat on low for 1 - 4 hours. Do not allow oil to go above 120 degrees. Strain and store in a clean dark glass bottle and label.

To make more potent medicinal oils, you can repeat the process by adding a fresh batch of herbs to the strained oil and infuse for 2 more weeks and strain again. Store the oils in a cool, dark location for up to a year. If storing for longer periods of time refrigeration may be necessary to keep oils from going rancid. Make sure bottles and all equipment are very clean and dry. Remember to label everything with ingredients (name of herb and oil) and date!

Salves

Salves and ointments are made by the addition of a solidifying agent like beeswax or cocoa butter to an infused oil. The general rule is ½ - 1oz solidifying agent to 1 cup oil but this is not an exact science and firmness is determined by personal preference and experimentation. There are many resources with recipes for great salves. See some recommendations in the **Resources**.

Appendix

Creams

Creams consist of infused herbal oils, solidifying agents and water or herbal tea. The creamy consistency is created by blending the ingredients till emulsified. They are a little trickier to make, but fun and it is so nice to have skin products made with simple ingredients that are healing, of good quality and infused with your personal energy.

Tinctures

Tinctures are herbs soaked in alcohol. They allow one to make a large quantity of herbal remedy and store it for several years. They are effective in very small amounts because they are so concentrated and extract constituents that water cannot. They are usually taken internally a dropperful at a time, but can also be added to teas or baths (for topical use). Tinctures are made by steeping fresh or dried herbs in alcohol. The alcohol extracts the volatile oils and active constituents from the herbs. Commercial tinctures are most often made with 100 proof alcohol (Everclear) and water. For the home kitchen, using vodka (80-85 proof, 40-45% alcohol) or brandy (74-80 proof, 30-40% alcohol) is fine when using dried herb. Fresh plant material contains water and a stronger solvent like 75-100% alcohol is preferred. There many books that provide the recommended alcohol ratios for specific plants (see **Resources**). Tinctures with at least 25% alcohol (50 proof) content will keep indefinitely.

The traditional folk method is to fill a jar with the cut up herb and cover with the alcohol of your choice to fully immerse the herb. Use a clean glass jar, and be sure to label it with the ingredients and date. Let steep at room temperature, shaking the jar every couple of days. After 2-6 weeks, strain the mixture through a fine cloth, paper coffee filter, or cheesecloth. The plant material can be added to the compost. Pour the tincture into clean, dark bottles, label and store out of the sun until needed.

Vinegars

A mineral-rich extract can be made by preserving herbs in organic apple cider vinegar in the same manner as making a tincture. Place fresh or dried herb in a clean glass jar and completely cover with the vinegar. Let stand for 2-6 weeks, shaking frequently. Strain and store in a clean bottle. These should last about a year. This can be used in cooking and salad dressings. The vinegar can also be used externally as a skin toner, in the bath or as a hair rinse. It should be diluted for external use.

Liniments

Topical liniments are made in the same manner as alcohol tinctures and vinegars. They are only used on the skin and never taken internally. They are used to soothe sore muscles, bruises and sprains and as a disinfectant for wounds. You can use a grain alcohol and pure water mixture, rubbing alcohol, witch hazel, vodka or vinegar to infuse the herbs. Allow it to steep for 2 to 6 weeks, shaking regularly, strain, label and store in a clean dark glass bottle. Be sure to label for EXTERNAL USE ONLY. Liniments have an indefinite shelf life.

Appendix

Edible Preparations

Following is a simplified list of a few ways to prepare the fruits (leaves, tubers, etc.) of your labors. Please see the **Resources** for several excellent cookbooks for further inspiration.

Raw plants

Knowing what plants, and plant parts, to use raw can lead to wonderful trail munchies! Probably out of all the methods listed here, using raw plants requires the most dedicated care to make sure you have the correct plant and plant part. Many times heat will destroy toxins and collecting errors can be made with minimal negative effects. Try and collect young and tender parts. Use fresh plants in salads, as garnishes, on sandwiches (great while hiking!), and in stir fries. Mix strong tasting plants with milder plants (unless, of course, you like it HOT!). Also be diligent about collecting guidelines to avoid contaminants. Never, ever eat plant parts raw from plants growing in or around water. The abundance of nasty water bugs could lead to nasty surprises.

Drying

Drying is one of the oldest methods of preserving fruits, roots and leaves. It is simply passing warm air (or using the sun's heat energy) over prepared foods to remove the moisture and can be accomplished with home dehydrators, an oven set to the lowest setting or the sun's heat. Use trays that will allow maximum air flow. Dehydrators come with adequate trays, for the oven I have used window screens, covered with cheesecloth, stretched over cake-cooling racks. If you are drying fruit leather use solid drying racks. When drying in an oven make sure the door is propped open slightly and the temperature is set to the lowest setting (maximum temperature of 140F). High temperatures can actually change the flavor and texture of many plant parts.

Prepare all plants by washing them thoroughly and slicing the larger plant parts. Fruit can generally be placed directly in the dehydrator or oven, whereas vegetables generally need to be blanched or parboiled before dehydrating. Again, check with recipe books for specific plants. Make sure the plant parts are separated on the drying racks to increase air flow and turn the pieces occasionally during the drying period.

Use the same tray types for drying in the sun, bringing the trays indoors each night so moisture does not settle on the plant parts. Do not dry plants outside if there is air pollution. Check the pieces often and turn if necessary.

To check for dryness, remove a piece or two from the rack and let cool completely to judge for doneness. If drying fruit leather, let the whole rack cool slightly before testing for dryness. Store the completely dried parts in clean glass jars and seal tightly. Make sure to check the jars over the next several days for moisture on the glass. You will need to dry the food a little longer if moisture is noticed. Dehydrated foods can be stored in clean glass jars for up to a year. If you notice any mold, discard it without tasting it. Fruits can generally be enjoyed directly from the jar, but roots and other hefty plant parts should be soaked first to rehydrate them before being used.

We can't leave out the leaves! Many plant leaves will maintain much of their aromas and flavors, as well as their nutrition, if they are simply bundled on the stem and hung in

Appendix

a warm, dry place to dry completely. Again, store the dried leaves in a sealed clean glass container.

Boiling

Often, boiling is used to remove undesirable compounds such as bitter chemicals or toxins. Boiling also breaks down plant fibers of many tough plant parts, making them more palatable. Sometimes a pinch of baking soda can be added to really tough plants to make them more tender. If the plant being boiled is not extremely bitter or contains toxins, then often the remaining water can be used in soups or drunk. When boiling plants with poisons, make sure to discard the water.

Baking and Roasting

Like boiling, baking or roasting can often improve the texture and palatability of a plant. Many times, this type of cooking breaks down somewhat indigestible carbohydrates, producing a food that is easier to digest because of the higher sugar content.

Syrups

Adding some form of sugar to a fruit juice, heating the mixture until the sugar dissolves and the juice comes to a boil, then canning the liquid is a simple way to add a hint of fruit (and sometimes leaves (mints) and roots (sweet anise)) to make a syrup that you can use as maple syrup, in beverages, or over ice cream. Teresa Marrone has a good recipe in her book *Cooking with Wild Berries and Fruits*.

Sap from aspen trees was considered a delicacy by the Ute people.

Preserving and Canning

We usually think of jams and jellies when we think of canning, and the recipes for our yummy fruits are numerous. But don't forget that we can pickle many items with different spices and produce some wonderful foods. Cattail shoots make excellent pickled foods.

Flours

Flours can be a time consuming project as the plant parts need to be cleaned then thoroughly dried before the plant is ground into flour or a coarser meal. Starchy fruits (e.g. grass grains and acorns) and the variety of root parts (rootstalks, tubers, bulbs and corms) are typically made into flours because of their higher starch content.

Freezing

If you have space in your freezer, freezing fleshy, sweet fruits such as raspberries, currants, serviceberries and strawberries is easy and practical. Wash the fruit well and remove any stems or leaves, blot excess water and place fruit in a single layer on a baking sheet and freeze overnight. Pack in containers in premeasured amounts for super easy use all through the winter. Or, after cleaning the fruit, simply pack it in premeasured amounts in containers and freeze. I prefer the beautifully preserved fruit from the baking sheet, but either method works well.

Appendix

Teas and other drinks

Teas made from leaves, flowers, bark or fruit are a delicious alternative to sugar-laden and caffeinated drinks. Simply pour boiling water over the plant parts and let them steep for 10 minutes or so. The longer they are steeped, the stronger the beverage. You will find plants or plant mixes that appeal to your taste buds. The strength of the tea can be modified by increasing or decreasing the amount of plant material used and/or the length of the steeping. In summer many of these teas are refreshing cold drinks. Add a few berries or other natural sweeteners and enjoy!

Parching

Parching is sometimes used to improve the flavor of certain grains or nutlets. Lightly roast the dried seeds or nut-like fruits at a low temperature, stirring often, until they are a golden brown. Use parching to break open tough seed coats such as those on yellow pond lily, which will open creating unique 'popcorn.'

Fermentation

This is another ancient method of preserving a variety of plant parts using microbial fermentation to produce beers, wines and vinegars. There are many books reviewing the processes and details of fermentation; it is a time-consuming but extremely rewarding process.

Appendix

A Few Yummy Recipes

Serviceberry Catsup

1½ quarts fresh serviceberry pulp
1 pint vinegar
1 Tbsp. ground cinnamon
1 Tbsp. ground cloves
1 Tbsp. allspice
Dash cayenne pepper
4 cups sugar
½ Tbsp. salt
With potato masher, mash enough serviceberries to make 1½ quarts of pulp. Mix with remaining ingredients and cook over slow heat until mixture thickens. Store in refrigerator in covered jars or freeze. Good with game.

— *The Rocky Mountain Wild Foods Cookbook*

Golden Cattail Biscuits

1 cup flour
1 cup cattail pollen, cleaned
2 tsp. baking powder
½ tsp. salt
4 Tbsp. butter, melted
¼ cup yogurt
Mix the dry ingredients together. Add the butter and yogurt and mix well. Roll out the dough on a floured surface. Cut into biscuits and bake on a greased pan at 450F for 12-15 minutes or until done.

— *A Taste of Nature: Edible Plants of the Southwest and How to Prepare Them*

Illustrated Glossary

Flower structure

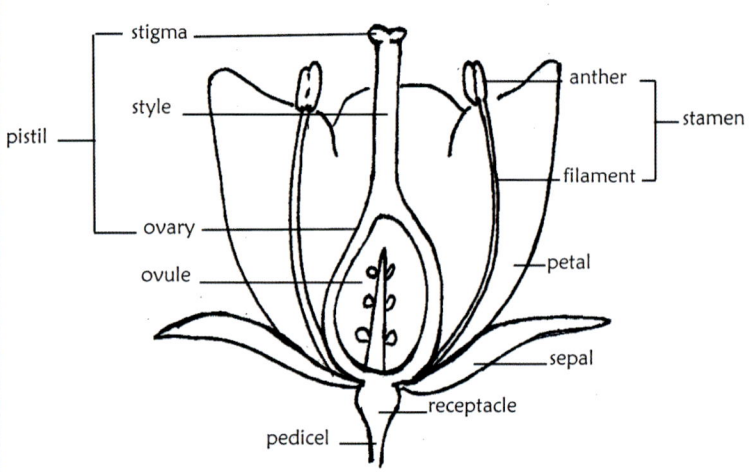

Inflorescences

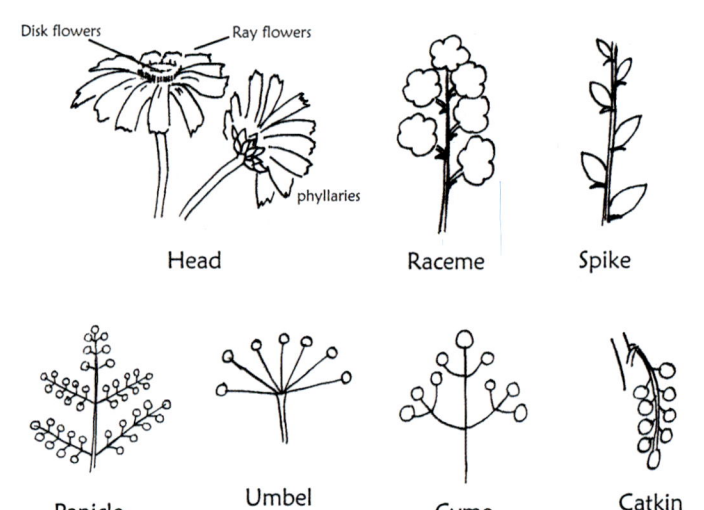

Illustrated Glossary

Leaf arrangement

Alternate　　Opposite　　Whorled　　Basal

Leaf structure

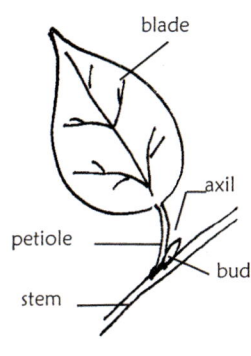 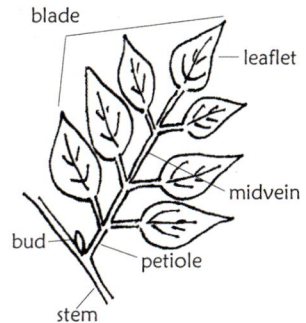

Simple leaf　　　　　　Compound leaf

Palmately compound　　Pinnately compound

Leaf shapes

Filliform　　Linear　　Lanceolate　　Ovate

Glossary

GLOSSARY OF BOTANICAL TERMS

achene: a small dry fruit that does not split open, often seed-like

alternate: found singly at each node or joint (like leaves on a stem), or regularly between other organs (like stamens alternate with petals)

annual: a plant completing its life cycle in one season

anther: the male, pollen-bearing part of the flower

axil: the position between a side organ (like a leaf) and the area to which it is attached (like the stem)

biennial: living for two years, usually producing flowers and seeds in the second year

bract: a specialized leaf with a flower arising from its axil

bulb: a short, vertical underground stem with thickened leaves or leaf bases

bulblet: a small bulb-like structure produced in a leaf axil or replacing a flower

calyx: the outermost set of floral parts called sepals collectively

catkin: a dense spike or raceme of many unisexual, naked flowers that lack petals and sepals but have a bract

clasping: embracing or surrounding, usually in reference to a leaf base around a stem

corm: swollen stem base containing food material and bearing buds in the axils of the scale-like remains of leaves from the previous season

corolla: the second circle of floral parts, the petals collectively

disc (or disk) florets: a small, tubular flower in a flowerhead of the Aster Family, usually clustered at the center of the head

filament: the stalk or stem supporting the anther

floret: a small flower, usually one of several in a cluster

follicle: a dry, pod-like fruit, splitting open along a single line on one side

head: a cluster of flowers crowding the tip of the stem

inflorescence: the arrangement of flowers on the stem

involucre: a set of bracts closely associated with one another, encircling and immediately below a flower cluster

node: a point on the stem where leaves or branches originate

nutlet: a small, hard, dry 1-seeded fruit or part of a fruit, not splitting open

ovary, ovule: ovules, unfertilized seeds, are found inside the ovary. After ovules are fertilized, the ovary develops into a fruit or capsule, and the ovules become seeds.

panicle: a loosely branched cluster of stalked flowers or spikelets, blooming from the bottom up

pappus: a modified calyx forming a crown of bristles, hairs, awns, teeth or scales at the tip of the seed

Glossary

petiole: the stem supporting a leaf

perennial: living for 3 or more years, usually flowering and fruiting for many years

petals: the floral leaves inside the sepals, usually colored or white

pinnate: with branches, lobes, leaflets or veins arranged on both sides of a central stalk or vein (like a feather)

pistil: the female parts collectively

raceme: an unbranched cluster of stalked flowers on a common, elongated central stalk, blooming from the bottom up

ray floret: a small, flattened, strap-like flower typically in the Aster Family, usually radiating from the edge of the flower

rhizome: a horizontal underground stem

rosette: a cluster of crowded, usually basal leaves in a circular arrangement

sepals: the outermost floral leaf, usually green and surrounding the flower in bud

serrate: having sharp short teeth along the edge of a leaf

sessile: lacking a stem or pedicle, attached at the base

spike: a simple, unbranched flower cluster with stalkless flowers arranged along a single axis

spur: a hollow appendage on a petal or sepal

stamen: the male flower parts collectively

stigma: the pollen-receiving part of the flower, typically very sticky

stipule: a pair of bract-like or leaf-like appendages at the base of the leaf stalk

stolon: a slender, prostrate, spreading branch, rooting and often developing new shoots and/or plants at its nodes or tip

style: the part of the pistil connecting the stigma to the ovary, often elongated and stalk-like

taproot: a root system with a prominent main root, directed vertically down, with smaller rootlets branching off, sometimes swollen and containing stored food

tendril: a slender, clasping or twining growth from a stem or leaf

tepals: showy flower parts when not differentiated into petals or sepals

tuber: a thick, creeping underground stem or root, serving for food storage and sometimes propagation

umbel: a round or flat-topped flower cluster in which all flower stalks are of similar length and arise from the same point

whorl: a ring of 3 or more similar structures (like leaves, branches or flowers) arising from one node

Glossary

GLOSSARY OF MEDICAL TERMS

adaptogen: a substance that increases the body's ability to return to normal in the presence of stress. Generally works by strengthening the immune, nervous, or glandular system.

alkaloid: any of a large class of nitrogen containing organic compounds found especially in seed plants. These include codeine and morphine and have many medicinal applications. Usually bitter tasting.

alterative: a substance that gradually alters or changes a condition. It helps restore proper body function. Often, it is a medicine that cures an illness by gradually restoring health.

analgesic: a substance that relieves pain

anodyne: relieves pain; is milder than an analgesic

antibacterial: kills bacteria

antibiotic: an agent that kills or inhibits growth or multiplication of a living organism, especially bacteria or other microorganisms

antiallergic: preventing or relieving allergies

anticatarrhal: a substance that removes excess mucus in sinus and other areas

antidandruff: treats dandruff

anthelmintic: an agent that expels parasitic worms from the stomach or intestines

antifungal: an agent that kills or inhibits fungi

anti-inflammatory: a substance that soothes or reduces tissue inflammation

antimicrobial: an agent that kills or inhibits microorganisms

antipyretic: a substance which dissipates a fever

antirheumatic: a preparation that eases the discomfort of rheumatism, a condition that causes inflammation and pain in the joints and muscles

antiseptic: prevents sepsis, decay, putrefaction; an agent that kills microbes

antispasmodic: a substance that will relieve or prevent muscular or nervous spasms and cramps

antiviral: a substance that inhibits the proliferation and viability of infectious viruses

aromatherapy: the use of essential oils in the treatment of medical, emotional, spiritual and cosmetic problems

aromatic: having a strong, distinctive odor

astringent: an agent that causes the constriction of tissues, usually used externally to stop bleeding, secretions and surface inflammation and distension. Reducing irritation helps protect against infection. To be effective all astringents must come in direct contact with the inflamed or bleeding tissue to be effective.

bactericidal: an agent that kills bacteria

Glossary

bacteriostatic: stopping or slowing the proliferation of bacteria

bitter: a bitter tasting substance or formula that stimulates appetite, digestion and liver function

bronchodilating: an agent that dilates the bronchial tubes

cardiotonic: a tonic for the heart

carminative: a substance that soothes the digestive system by checking the formation of gas and helps to dispel whatever gas has already formed

catarrh: an inflammation of any mucous membrane especially that of the respiratory system that implies excess secretions and congestion

cathartic: a laxative or purgative, which causes evacuation of the bowels. A laxative is a gentle cathartic while a purgative is much more forceful.

cholagogue: an agent that promotes the discharge of bile from the gall bladder

counterirritant: a substance applied to the skin, like a liniment or mustard plaster, to produce an irritating, heating, or vasodilating effect, in order to speed local healing by increasing circulation of the blood and radiating the heat inward to inflamed tissues

decoction: a preparation made by simmering plant material in water for an extended period. Usually it is made of hard plant substances such as roots, bark, or seeds.

demulcent: an agent, usually oily or mucilaginous, that soothes internal membranes providing a protective coating and calming irritation

depurative: eliminates toxins and purifies the system, especially the blood

diaphoretic: a substance taken internally to promote sweating either by dilating the peripheral blood vessels or directly stimulating the nerves that affect sweat glands

digestive: an agent that promotes digestion

diuretic: a substance that acts on the kidneys to increase the flow of urine

emetic: an agent which induces vomiting

emmenagogue: tones the female reproductive system and promotes menstruation

emollient: a substance which, when applied externally, softens and soothes the skin

expectorant: a substance that stimulates the outflow of mucus from the lungs and bronchial mucosa, usually a respiratory tonic

extract: a product made by removing the essential constituents from herbs through pressing, distilling, or using a solvent

febrifuge: reduces fevers

fungicide: an agent that kills fungi

Glossary

galactagogue or lactifuge: promotes the secretion of milk

hemostatic: any substance used to stop or slow bleeding

hepatic: any substance that affects the liver; usually refers to strengthening and toning

hypertensive: causing a rise in blood pressure

histamine: a defense substance present in all tissues of the body responsible for most inflammations

hypnotic: induces sleep (not a hypnotic trance)

hypoglycemic: causing a lowering of blood sugar

hypotensive: lowers abnormally high blood pressure

infusion: an extract of a substance made by soaking it in water. A tea is an infusion. This is the simplest, most common method for extracting the medicinal properties from plant materials. The fresh or dried herb is covered with boiling water and allowed to steep for 5 to 10 minutes.

laxative: a substance that stimulates bowel action

mucilaginous: plant constituent that has a mucous-like (slimy) feel when crushed; used to soothe inflammations

nervine: a substance that calms or quiets nervousness, tension or excitement. Usually it is a nervous system tonic, relaxant or stimulant

odontalgic: treats toothache and other problems of teeth and gums

oxytocic: stimulates contraction of the uterus and can help in childbirth

pectoral: an agent that relieves ailments of the chest and lungs

poultice: a remedy applied to the surface of the body usually made of fresh plant matter that has been crushed or soaked into a soft mass, then placed between two pieces of cloth for application to the skin

purgative: a strong cathartic, given to relieve severe constipation

rubefacient: a substance which, when rubbed into the skin, reddens the skin by attracting blood to the surface

sedative: a remedy that calms the nerves and promotes sleep

sialagogue: stimulate secretion of saliva

simple: a medicinal herb, working singly as a complete form of treatment. They are usually locally grown or indigenous plants and usually mild

soporific: a substance that induces sleep

stimulant: an agent that causes increased activity of another agent, cell, tissue, organ, or organism

Glossary

stomachic: a substance that counteracts or relieves cramps and aids and improves the action of the stomach

styptic: an agent that checks bleeding by contracting the blood vessels

tincture: a solution of plant material in alcohol

tonic: a substance that invigorates or strengthens the system. Often taken to strengthen and prevent disease, especially chronic disease. They tend to stimulate deficient functions and act as an alterative. Slower acting than a stimulant, bringing steady improvement.

vasoconstrictor: an agent that narrows blood vessel openings, restricting the flow of blood through them

vasodilator: an agent that expands and relaxes blood vessels

vermifuge: a medicine that destroys intestinal parasites and helps expel them

vulnerary: a substance that soothes or heals wounds and sores. Can be an antibiotic, antiseptic, styptic or heal through cell regeneration.

Resources

Anderson, M. Kat. *Tending the Wild.* University of California Press, 2006.

Anderson, Richard M., Jay Dee Gunnell and Jerry L. Goodspeed. *Wildflowers of the Mountain West.* Utah State University Press, 2012.

Beidleman, Linda H., Richard G. Beidleman and Beatrice E. Willard. *Plants of Rocky Mountain National Park.* Falcon Publishing, 2000.

Benedict, Audrey DeLella. *Naturalist's Guide to the Southern Rockies.* Fulcrum Publishing, 2008.

Blair, Katrina. *Local Wild Life.* Turtle Publications, 2009.

Brill, "Wildman" Steve and Dean Evelyn. *Identifying and Harvesting Edible and Medicinal Plants in Wild (and not so wild) Places.* William Morrow Paperbacks, 1994.

Buhner, Stephen Harrod. T*he Lost Language of Plants: The Ecological Importance of Plant Medicines for Life on Earth.* Chelsea Green Publishing, 2002.

Buhner, Stephen Harrod. *The Secret Teachings of Plants: The Intelligence of the Heart in the Direct Perception of Nature.* Bear & Company, 2004.

Buhner, Stephen Harrod. *Herbal Antibiotics: Natural Alternatives for Treating Drug-Resistant Bacteria.* 2nd Edition. Storey Publishing, 2012.

Burrill, Larry C., Steven A. Dewey and David W. Cudney, et.al. *Weeds of the West.* Western Society of Weed Science, 1996.

Busco, Jane and Nancy R. Morin. *Native Plants for High Elevation Western Gardens.* Fulcrum Publishing, 2003.

Carter, Jack L. *Trees and Shrubs of Colorado.* 2nd ed. Mimbres Publishing, 2006.

Corbridge, James N. and William A. Weber. *A Rocky Mountain Lichen Primer.* University Press of Colorado, 1998.

Cutts, Gretchen S. *Potions, Portions, Poisons, Indian and Settler Plant Uses.* Gretchen Cutts, 1976.

Darrow Warren, Kathy. *Wild About Wildflowers, Extreme Botanizing in Crested Butte, Wildflower Capital of Colorado.* Fort Collins: Wildkat Publishing Co., 2006.

DeLella Benedict, Audrey. *The Naturalist's Guide to the Southern Rockies.* Fulcrum, 2008

Densmore, Frances. *How Indians Use Wild Plants for Food, Medicine and Crafts.* Dover Publications Inc., 1974.

Derig, Betty B. and Margaret C. Fuller. *Wild Berries of the West.* Mountain Press Publishing Co., 2001.

Ellefson, Connie Lockhart and David Winger. *Xeriscape Colorado: The Complete Guide.* Westcliffe Publishers, 2004.

Elpel, Thomas J. *Botany in a Day: Thomas J. Elpel's Herbal Field Guide to Plant Families,* 4th Ed. Hops Press, 2000.

Elpel, Thomas J. and Reed, Kris. *Foraging the Mountain West.* HOPS Press LLC, 2014.

Farnsworth, Kahanah. *A Taste of Nature: Edible Plants of the Southwest and How to Prepare Them.* Ancient City Press, 1997

Forrest, Jessica; Inouye, David W.; Thomson, James D. *"Flowering phenology in subalpine meadows; Does climate variation influence community co-flowering patterns?"* Ecology 91(2), 2010, pp.431-440, Ecological Society of America.

Gershuny, Grace. *Start with the Soil: The Organic Gardener's Guide to Improving Soil for Higher Yields, More Beautiful Flowers, and a Healthy, Easy-Care Garden.* Rodale Press, 1993.

Resources

Gladstar, Rosemary. *Herbal Healing for Women*. Touchstone, 1993.

Gladstar, Rosemary. *Herbal Recipes for Vibrant Health: 175 Teas, Tonics, Oils, Salves, Tinctures, and Other Natural Remedies for the Entire Family*. Storey Publishing, 2008

Gladstar, Rosemary. *Rosemary Gladstar's Medicinal Herbs: A Beginner's Guide: 33 Healing Herbs to Know, Grow, and Use*. Storey Publishing, 2012.

Grant, M. & Mitton, J. (2010) *Case Study: The Glorious, Golden, and Gigantic Quaking Aspen*. Nature Education Knowledge 3(10):40

Green, James. *The Herbal Medicine-Maker's Handbook: A Home Manual*. Crossing Press, 2000.

Harrington, H.D. *Edible Native Plants of the Rocky Mountains*. The University of New Mexico Press, 1967.

Hartung, Tammi. *Homegrown Herbs: A Complete Guide to Growing, Using, and Enjoying More than 100 Herbs*. Storey Publishing, 2011.

Hartung, Tammi. *Growing 101 Herbs that Heal*, Storey Publishing, 2000.

Hemenway, Toby. *Gaia's Garden, Second Edition: A Guide to Home-scale Permaculture*. Chelsea Green Publishing, 2009.

Hobbs, Christopher and Steven Foster. *Western Medicinal Plants and Herbs*. Houghton Mifflin Harcourt, 2002.

Huggins, Lindsey. Janis *Snowmass Village Wild at Heart,* People's Press, 2008

Kershaw, Linda, Andy MacKinnon and Jim Pojar. *Plants of the Rocky Mountains*. Lone Pine Publishing, 1998.

Kershaw, Linda. *Edible and Medicinal Plants of the Rockies*. Lone Pine Publishing, 2000.

Kindscher, Kelly. *Edible Wild Plants of the Prairie: An Ethnobotanical Guide*. University of Kansas Press, 1987.

Kindscher, Kelly. *Medicinal Wild Plants of the Prairie: An Ethnobotanical Guide*. University of Kansas Press, 1992.

Kirk, Donald R. *Wild Edible Plants of Western North America*. Naturegraph Publishers, 1975.

Komarek, Susan. *Flora of the San Juans: A Field Guide to the Mountain Plants of Southwest Colorado*. Kivaki Press, 1995.

Kress, Henriette. *Practical Herbs. Herbs and Herbal Therapy.* Henriette Kress, 2011.

Krumm, Bob. *The Rocky Mountain Berry Book*. Falcon Press, 1991.

Marrone, Teresa. *Cooking with Wild Berries and Fruits of the Rocky Mountain States*. Adventure Publications Inc. 2012.

Marrone, Teresa. *Wild Berries and Fruits Field Guide; Rocky Mountain States*. Adventure Publications Inc., 2012.

McBeth, Sally. *Ethnographic Overview Draft #2*. Colorado National Monument. 2010.

Moerman, Daniel E. *Native American Ethnobotany*. Timber Press, 1998.

Montgomery, Pam. *Plant Spirit Healing: A Guide to Working with Plant Consciousness*. Bear & Company, 2008.

Moore, Michael. *Medicinal Plants of the Mountain West*. Museum of New Mexico Press, 2003.

Morgan, Lisbeth. *Foraging the Rocky Mountains: Finding, Identifying, and Preparing Edible Wild Foods In The Rockies* (Foraging Series). Falcon Guides, 2013.

Nyerges, Christopher. *Guide to Wild Foods and Useful Plants*. Chicago Review Press.

Resources

Pesman, M. Walker. *Meet the Natives: A Field Guide to Rocky Mountain Wildflowers, Trees and Shrubs.* Johnson Books, 2012.

Pfaffman, Garrick. *Rocky Mountain Plants* (Family Field Guides, Book 2), BearBop Press, 2007.

Phillips, H. Wayne. *Central Rocky Mountain Wildflowers.* Globe Pequot Press, 1999.

Robertson, Leigh. *Southern Rocky Mountain Wildflowers.* Falcon Publishing, 1999.

Sanders, Jack. *The Secrets of Wildflowers.* The Lyons Press, 2003.

Scott, Timothy Lee. *Invasive Plant Medicine: The Ecological Benefits and Healing Abilities of Invasives.* Healing Arts Press, 2010.

Seebeck, Cattail Bob. *Best Tasting Wild Plants of Colorado and the Rockies.* Westcliffe Publishers, 1998.

Smith, Anne M. *Ethnography of the Northern Utes.* Museum of New Mexico, 1974.

Sweet, Muriel. *Common Edible and Useful Plants of the West.* Naturegraph Publishers, 1976.

Taylor, Ronald J. *Sagebrush Country, A Wildflower Sanctuary.* Mountain Press Publishing Co., 1992.

Thayer, Samuel. *Nature's Garden: A Guide to Identifying, Harvesting, and Preparing Edible Wild Plants.* Forager's Harvest Press, 2010.

Tilford, Gregory L. *Edible and Medicinal Plants of the West.* Mountain Press Publishing Co., 1997.

Toensmeier, Eric. *Indigenous Use and Management of Native Plants of the Rocky Mountain and Prairie Regions.* 2010.

Turner, Nancy J. and von Aderkas, Patrick. *The North American Guide to Common Poisonous Plants and Mushrooms.* Timber Press, 2009.

Van Allen Murphey, Edith. *Indian Uses of Native Plants.* Meyerbooks, 1990.

Weber, William A. and Wittman, Ronald C. *Colorado Flora: Western Slope.* 4th ed. University Press of Colorado, 2012.

Weed, Susan. *Healing Wise* (Wise Woman Healing Series). Ash Tree Publishing, 2003.

Wiles, Briana. *Mountain States Medicinal Plants*, Timber Press, 2018

Wiles, Briana. *Mountain States Foraging*, Timber Press 2016

Willard, Terry. *Edible and Medicinal Plants of the Rocky Mountains and Neighbouring Territories.* Wild Rose College of Natural Healing, 1992.

Williamson, Darcy. *The Rocky Mountain Wild Foods Cookbook.* The Caxton Press, 1995.

Wingate, Janet L. *Rocky Mountain Flower Finder,* Nature Study Guild Publishers, 1990.

Resources

WEBLIOGRAPHY

Biota of North America. The. http://bonap.org/
BONAP's North American Plant Atlas. http://bonap.net/napa
Colorado Native Plant Society. http://www.conps.org
Colorado Natural Heritage Program. http://www.cnhp.colostate.edu
Colorado Plant Database. http://jeffco.us/coopext/intro.jsp
Great Basin Natives. http://www.greatbasinnatives.com
Henriette's Herbal. http://www.henriettesherbal.com
HerbWorld Online. http://www.herbworld.com
Integrated Taxonomic Information System. http://www.itis.gov/
Learning Herbs.com. http://www.learningherbs.com
Moore, Michael, Southwest School of Botanical Medicine.
    http://www.swsbm.com
Native American Ethnobotany. University of Michigan – Dearborn.
    http://herb.umd.umich.edu/
Plant Select. http://www.plantselect.com
Plants of the Southwest. http://www.plantsofthesouthwest.com
Plants for a Future. http://www.pfaf.org/user/default.aspx
Rocky Mountain Herbarium. http://www.rmh.uwyo.edu
Rose, Kiva, The Medicine Woman's Roots. http://bearmedicineherbals.com
Southwest Colorado Wildflowers. http://www.swcoloradowildflowers.com
United Plant Savers. http://www.unitedplantsavers.org/
USDA Plant Database. http://www.plants.usda.gov
"Wildman": Steve Brill. http://www.wildmanstevebrill.com/
Wildfood Girl. http://wildfoodgirl.com
Weed, Susan. http://www.susunweed.com
Western Native Seed. http://www.westernnativeseed.com
Xeric Gardener, The. http://www.waterwisegardening.com
X-Rated Gardening. http://www.coloradenclub.com

Index to Common and Scientific Names

Abies lasiocarpa var. *lasiocarpa* 24
Achillea millefolium 90
Aconitum columbianum 230
Actaea rubra 230
Agastache urticifolia 154
alfalfa 142
Allium 76
    A. brandegeei
    A. cernuum
    A. geyeri
Amelanchier alnifolia 58
Anaphalis margaritacea 92
Anemone multifida var. *multifida* 244
Antennaria spp. 94
Apocynum androsaemifolium 214
Aquilegia 232
    A. coerulea
    A. elegantula
Arctostaphylos uva-ursi 44
Arnica spp. 218
arnicas 218
Artemisia spp.
    A. frigida 96
    A. ludoviciana 96
    A. scopulorum 96
    A. cana ssp. *cana* 36
    A. nova 36
    A. tridentata ssp. *vaseyana* 36
Asclepias speciosa 88
aspen 72
Astragalus spp. 226
balsamroot, arrowleaf 98
Balsamorhiza sagittata 98
baneberry, western 230
Barbarea orthocerus 128
bedstraw, northern 192
bee balm 158
bearberry 44
bergamot, wild 158
bilberry 46
biscuitroot 80
    fernleaf
    Gray's
    Wasatch desert parsley
bistort 180
Bistorta bistortoides 180
bittercress 126

buckwheat 182
    false
    subalpine
    sulphur flower
buffaloberry 42
bunny in the grass 194
buttercups 242
Calochortus gunnisonii 130
Caltha leptosepala 152
Campanula 132
    C. rotundifolia
    C. parryi
candytuft, wild 126
Capsella bursa-pastoris 122
Cardamine cordifolia 126
Castilleja spp. 240
cattail, broad-leaved 198
Ceanothus velutinus 56
Chamerion angustifolium subsp.
    circumvagum 168
Chenopodium album 136
chicory 100
Chimaphila umbellata
    ssp. *occidentalis* 186
chokecherry 62
Chrysothamnus viscidiflorus spp.
    viscidiflous 34
Cichorium intybus 100
Cicuta maculata 216
cinquefoils 188
cleavers 192
Clematis 244
    C. occidentalis var. *occidentalis*
    C. hirsutissima var. *hirsutissima*
clover, red 144
columbine 232
    Colorado blue
    western red
Conium maculatum 216
corn lily 238
Cornus sericea ssp. *sericea* 40
cow parsnip 78
cranesbill 150
Crataegus 60
    C. erythropoda
    C. rivularis
cress, hoary 126

Index

currant 52
    golden
    red prickly
    wax
Cynoglossum officinale 120
dandelion 114
death camas 238
    meadow
    mountain
Delphinium 234
    D. barbeyi
    D. glaucum
    D. nuttallianum
    D. ramosum
Descurainia sophia 128
dock 184
    yellow or curly
dogbane, spreading 214
dogwood, red-osier 40
Douglas-fir, Rocky Mountain 30
elderberry, red 220
elkbrush 56
Equisetum spp. 208
Ericameria 34
    E. parryi var. *parryi*
    E. nauseosa var. *nauseosa*
Eriogonum 182
    E. umbellatum
    E. umbellatum var. *majus*
Erythronium grandiflorum 162
evening primrose 170
    stemless white
    yellow
Fragaria 190
    F. vesca ssp. *bracteata*
    F. virginiana ssp. *glauca*
false hellebore 238
figwort 194
fireweed 168
flax, wild blue 164
Frasera speciosa 148
Galium 192
    G. boreale
    G. aparine
gentian
    bottle 146
    fringed 146
    green 148

Gentiana parryi 146
Gentianopsis thermalis 146
geranium 150
    Richardson's
    sticky
Geranium 150
    G. richardsonii
    G. viscosissimum
glacier lily 162
goldenbanner, mountain 222
goldenglow 110
goldenrods 112
gooseberries 54
Grindelia squarrosa 102
grouse whortleberry 46
gumweed 102
harebell 132
    common
    parry's bellflower
hawthorn 60
    cerro
    river
Helianthella 104
    H. quinquenervis
    H. uniflora
hemlock 216
    poison
    spotted-water
Heracleum sphondyllium
    ssp. *montanum* 78
hops 134
horsemint, nettle leaf 154
horsetail 208
hound's tongue 120
huckleberry 46
    broom
    dwarf
    low
Humulus lupulus var. *neomexicanus* 134
hyssop, giant 154
iris, Rocky Mountain 236
Iris missouriensis 236
junipers 22
    common
    Rocky Mountain
Juniperus 22
    J. communis ssp. *montana*
    J. scopulorum

Index

kinnikinnick 44
Lactuca serriola 106
lamb's-quarters 136
larkspur 234
    mountain
    Nuttall's
    sierra
    subalpine
Lathyrus lanszwertii var. *leucanthus* 224
lettuce 106
    wild
    prickly
Lepidium draba 126
Ligusticum porteri 84
lily
    avalanche 162
    corn 238
    glacier 162
    mariposa 130
    sego 130
    yellow pond 166
lily of the valley, false 138
Linum lewisii var. *lewisii* 164
Lithospermum ruderale 222
locoweeds 226
Lomatium 80
    L. dissectum
    L bicolor var. *leptocarpum*
    L. grayi
lousewort 172-174
    elephant head 172
    fernleaf 172
    giant 172
    parrot's beak 174
lovage, Porter's 84
lupines 228
Lupinus spp. 228
Mahonia repens 38
Maianthemum 138
    M. racemosum ssp. *amplexicaule*
    M. stellatum
marsh-marigold 152
Matricaria discoidea 108
Medicago sativa 142
Mentha arvensis 156
milk vetch 226
    alpine
    Rocky Mountain

milkweed, showy 88
mint, field 156
Monarda fistulosa 158
monkshood 230
monument plant 148
mountain ash 70
mountain laurel 56
mule's ears 118
mullein 196
Mustard Family 122-129
    bittercress 126
    candytuft, wild 126
    cress, hoary 126
    mustard, tansy 128
    pennycress 124
    shepherd's purse 122
    wintercress 126
nettle, stinging 202
New Jersey tea 56
Noccaea fendleri ssp. *glauca* 126
Nuphar polysepala 166
Oak 48
    gambel
    scrub
Oenothera 170
    O. cespitosa
    O. elata ssp. *hirsutissima*
onion 76
    Brandegee's
    Geyer's
    nodding
Oregon grape 38
osha 84
Osmorhiza occidentalis 82
Oxytropis spp. 226
paintbrushes 240
pasqueflower 242
Pea Family 222-229
    goldenbanner, mountain 222
    locoweeds 226
    lupines 228
    milk vetches 226
    peavine, white-flowered 224
    vetch, American 224
pearly everlasting 92
peavine, white-flowered 224

Index

Pedicularis 172-175
    P. bracteosa 172
    P. grayi 172
    P. groenlandica 172
    P. racemosa 174
pennycress 124
Perideridia gairdneri 86
Picea 26
    P. engelmannii
    P. pungens
pigweed 136
pine, lodgepole 28
pineapple weed 108
Pinus contorta var. *latifolia* 28
pipsissewa 186
plantain 176
    common
    English
Plantago 176
    P. lanceolata
    P. major
poison ivy, western 214
poleo 156
Populus tremuloides 72
potentillas 188
Potentilla spp. 188
Prunella vulgaris 160
Prunus virginiana var. *demissa* 62
Pseudotsuga menziesii var. *glauca* 30
Pulsatilla patens ssp. *multifida* 242
pussytoes 94
quaking aspen 72
Quercus gambelii 48
rabbitbrush 34
    Parry's
    rubber
    yellow
ragwort 218
    arrowleaf
    tall
Ranunculus spp. 242
raspberry 68
red-osier dogwood 40
redroot 56
red willow 40

Ribes 52-55
    R. aureum 52
    R. cereum 54
    R. montigenum 54
Rosa woodsii 64
rose 64
    dog
    wild
    Wood's
Rubus 66-69
    R. idaeus 68
    R. parviflorus 66
Rudbeckia laciniata var. *ampla* 110
Rumex
    R. acetosella 178
    R. crispus 184
sage 96
    alpine
    fringed
    prairie
sagebrush 36
    black
    mountain
    silver
Salix spp. 74
salsify 116
    meadow
    yellow
Sambucus racemosa var. *racemosa* 220
saskatoon 58
Scrophularia lanceolata 194
scouring rush 208
Sedum lanceolatum var. *lanceolatum* 140
self-heal 160
Senecio 218
    S. serra
    S. triangularis
serviceberry 58
shepherd's purse 122
Shepherdia canadensis 42
snowberry, mountain 220
Solidago spp. 112
Solomon's seal 138
    false
    starry false
Sorbus scopulina 70

Index

sorrel, sheep 178
spruce 26
    Colorado blue
    Engelmann
stonecrop 140
stoneseed, western 220
strawberry 190
    wild
    woodland
subalpine fir 24
sugarbowls 244
sunflower, little 104
    five-nerved
    little
sweet anise 82
sweet root 82
Symphoricarpos rotundifolius 220
Taraxacum officinale 114
Thermopsis montana 222
thimbleberry 66
Thlaspi arvense 124
Toxicodendron rydbergii 214
Trifolium pratense 144
Tragopogon 116
    T. dubius
    T. pratensis
Typha latifolia 198
Urtica dioica 202
Usnea spp. 210
uva ursi 44
Vaccinium 46
    V. cespitosum
    V. myrtillus
    V. scoparium
Valeriana 204
    V. edulis
    V. occidentalis
valerian 204
    edible
    western
Veratrum californicum
    var. *californicum* 238
Verbascum thapsus 196
vetch, american 224
Vicia americana 224
Viola spp. 206

violets 206
virginsbower, western blue 244
whitetop 126
whortleberry 46
willow, red 40
willows 74
windflower 244
wintercress 126
Wyethia amplexicaulis 118
yampa 86
yarrow 90
yellow pond lily 166
Zigadenus 238
    Z. elegans ssp. *elegans*
    Z. venenosus

About the Authors

Karen Vail Mary O'Brien

MARY O'BRIEN has been studying and practicing herbal medicine since 1987. She studied with many renowned teachers including Rosemary Gladstar, Brigitte Mars, Stephen Buhner and Michael Moore and completed the Chartered Herbalist program with Dominion Herbal College. A Colorado native, she has lived in the Steamboat Springs area for over 30 years, where she teaches herbal medicine, leads medicinal plant walks, wildcrafts, makes herbal products and maintains a medicinal garden for Yampa River Botanic Park. Her journey has led her to also study the medicine wheel, sacred plant medicine, massage therapy, energy work and permaculture. Through her teaching, she emphasizes the use of native, non-native, and locally grown plants for plant-based self-care and incorporates permaculture principles. She is passionate about empowering others to recreate relationship with the plant world and become sustainable caretakers of Mother Earth.

KAREN VAIL received a B.S. in Horticulture, with a concentration in Botany from Colorado State University, and earned her Master's Degree in Gardening from the Royal Horticultural Society in England. She rode the "big yellow bus" for a year on Audubon Expedition Institute during her first year towards a Masters in Environmental Education, finishing up at Antioch in New England. Karen taught Botany at the local college and introduced people to local wildflowers through summer wildflower walks. Karen has published two books, *For the Joy of Wildflowers: Colorado Creations,* and *Yampa Valley Visions: Photography and Writing of a Yampa Valley Naturalist.* She has also co-published a video/DVD *A Season of Wildflowers: The Rocky Mountains.* She runs her own landscaping business, and continues to be active in education through a local environmental education organization, Yampatika, and a variety of local venues.